Springer Theses

Recognizing Outstanding Ph.D. Research

For further volumes:
http://www.springer.com/series/8790

Aims and Scope

The series "Springer Theses" brings together a selection of the very best Ph.D. theses from around the world and across the physical sciences. Nominated and endorsed by two recognized specialists, each published volume has been selected for its scientific excellence and the high impact of its contents for the pertinent field of research. For greater accessibility to non-specialists, the published versions include an extended introduction, as well as a foreword by the student's supervisor explaining the special relevance of the work for the field. As a whole, the series will provide a valuable resource both for newcomers to the research fields described, and for other scientists seeking detailed background information on special questions. Finally, it provides an accredited documentation of the valuable contributions made by today's younger generation of scientists.

Theses are accepted into the series by invited nomination only and must fulfill all of the following criteria

- They must be written in good English.
- The topic should fall within the confines of Chemistry, Physics, Earth Sciences and related interdisciplinary fields such as Materials, Nanoscience, Chemical Engineering, Complex Systems and Biophysics.
- The work reported in the thesis must represent a significant scientific advance.
- If the thesis includes previously published material, permission to reproduce this must be gained from the respective copyright holder.
- They must have been examined and passed during the 12 months prior to nomination.
- Each thesis should include a foreword by the supervisor outlining the significance of its content.
- The theses should have a clearly defined structure including an introduction accessible to scientists not expert in that particular field.

Yuan Li

Three Dimensional Solar Cells Based on Optical Confinement Geometries

Doctoral Thesis accepted by
Wake Forest University, USA

Springer

Author
Yuan Li
Department of Physics
Wake Forest University
Winston-Salem, NC, USA

Supervisor
David Carroll
Department of Physics
Wake Forest University
Winston-Salem, NC, USA

ISSN 2190-5053 ISSN 2190-5061 (electronic)
ISBN 978-1-4614-5698-8 ISBN 978-1-4614-5699-5 (eBook)
DOI 10.1007/978-1-4614-5699-5
Springer New York Heidelberg Dordrecht London

Library of Congress Control Number: 2012952095

Printed on acid-free paper

Springer is part of Springer Science+Business Media (www.springer.com)

Supervisor's Foreword

Three-dimensional (3D) optical geometries are becoming more and more common in the literature and lexicon of solar cells. Designed to improve the overall performance of a general absorber by employing optical confinement, the concept is meant to approach the thermodynamic limits allowed by the absorbers chosen. Currently there are several geometries under study, from spheres to boxes. In this work, the principles for fiber and tube geometries are worked out in detail. However, as we show through a combination of optical modeling as well as data directly from such devices, the principles uncovered are quite general and provide an important insight into the challenges of using such geometries with real, commercially viable absorbers.

Specifically, this work describes and reveals the basic operational nuances of 3D photovoltaics using three standard tools found in the field: equivalent circuit models, optical path ray tracing in the cavity, and absorber spectral response. These tools help us explain our experimentally observed absorption profiles and corresponding Jsc, Voc, filling factors, and EQE for such devices. In this work, they are applied to individual optical confinement geometry devices, integrated optical confinement geometry devices, and hybrid optical confinement geometry devices. Finally, we discuss the overall picture of these fascinating approaches to high-efficiency solar and point the way to their potential in solar farms, etc.

April 2012 David Carroll

Acknowledgements

First, I thank my supervisor Prof. David Carroll, who allowed me the freedom to pursue what interested me. During times when I would hover in a valley of scientific puzzle, he was always able to guide me through it. Without his support, I could not have completed and released my work of the past three and half years.

I would like to show my gratitude to Prof. Kim-Shapiro and Prof. Williams, who are my RAC committee. They gave me many great suggestions on my research. I am indebted to many people within Wake Forest University for making this project possible.

My wife Dan Xue, the most beautiful and kindest girl I have met, not only prepared delicious food for me, but also helped me with coding and web development. My biggest contribution in this work — a web-based photovoltaic analysis platform (OPVAP, www.opvap.com) — has mainly come from her assistance.

I would also like to thank the following members of Prof. Carroll's group for being excellent teammates: Wanyi, Wenxiao, Rob, Eric, Bob, Yonghua, and especially Huihui Huang and Mingjun Wang. If, say, Dr. Carroll guided me towards the direction of brightness, they are the persons who went through the valley with me. I especially appreciate the contribution and leadership of Mingjun on the work of tube-based DSSC, making this thesis more comprehensive in the study of hybrid 3D solar cell.

The collaborators Dewei Zhao, Xiang Wan, Junming Li, Pro f. Fujun Zhang, and Prof. Xiaowei Sun have provided assistance and support on many occasions. Xiang Wan provided strong support in Mathematics, which has been published in PRB.

Another friend, one who I must use a separate paragraph to express my thanks, is Dr. Xiao Xu. When I came to Wake Forest three years ago, he picked me up from the airport with another friend, and then, like a "Survival Manual," he introduced me to new friends, took me shopping and to a party, even gave suggestions for research and coding. I hope more contributions and collaborations will come from our friendship in the future.

Finally, I thank my parents. They have always supported my choices, and do their best to assist me in whatever problem I meet. Also, I want to extend thanks to all relatives and friends, past and current, for allowing me to finish my Ph.D smoothly.

Contents

Parts of this have been published in the following journal articles

H. Huang, Y. Li, M. Wang, W. Nie, W. Zhou, E. Peterson, J. Liu, G. Fang, D. Carroll, Photovoltaic-thermal solar energy collectors based on optical tubes. Sol. Energ. **85**, 450–454 (2011)

Y. Li, H. Huang, M. Wang, W. Nie, W. Huang, G. Fang, D. Carroll, Spectral response of fiber-based organic photovoltaics. Sol. Energ. Mat. Sol. C. (2011). doi:10.1016/j.solmat.2011.10.033

Y. Li, W. Nie, J. Liu, A. Partridge, and D. L. Carroll, The optics of organic photovoltaics: Fiber-based devices. IEEE J. Sel. Top. Quant. Electron. 1 (2010)

Y. Li, E.D. Peterson, H. Huang, M. Wang, D. Xue, W. Nie, W. Zhou, and D. L. Carroll, Tube-based geometries for organic photovoltaics. Appl. Phys. Lett. **96**, 243505 (2010)

Y. Li, M. Wang, H. Huang, W. Nie, Qi Li, E. D. Peterson, C. Robert, G. Fang, and D. L. Carroll, Influence on open-circuit voltage by optical heterogeneity in three-dimensional organic photovoltaics. Phys. Rev. B. (2011). doi: 10.1103/PhysRevB.84.085206

Y. Li, W. Zhou, D. Xue, J. Liu, E. Peterson, W. Nie, D. Carroll, Origins of performance in fiber-based organic photovoltaics. Appl. Phys. Lett. **95**, 203503 (2009)

W. Nie, Y. Li, W. Zhou, J. Liu, D. Carroll, Multi-layer deposition of conformal, transparent, conducting oxide films for device applications. Thin Solid Films. http://dx.doi.org/10.1016/j.tsf.2012.01.008

M. Wang, Y. Li, H. Huang, E. Peterson, W. Nie, W. Zeng, W. Huang, G. Fang, N. Sun, X. Zhao and D. Carroll, Thickness dependence of the MoO3 blocking layers on ZnO nanorod inverted organic photovoltaic devices. Appl. Phys. Lett. **98**, 103305 (2011)

Abbreviations and Synonyms

List of expressions which are used as synonyms throughout the text or relevant literature.

c	Light speed	m/s
Eff	Efficiency	%
E_g	Band gap	eV
FF	Fill factor	
h	Plank's constant	eV·s
J_e, J_h	Electron and hole current densities	mA/cm^2
J_{ph}	Photocurrent	mA/cm^2
J_{Rec}	Recombination current density	mA/cm^2
J_s	Reverse saturation current	mA/cm^2
J_{sc}	Short current density	mA/cm^2
k_B	Boltzmann constant	JK^{-1}
N_C, N_V	Effective densities of states in conduction and valence bands	1A/cm^2
n_D	Diode ideality factor	
$n_{e,}$ n_h	Concentration of electrons and holes	1/cm^3
q	Electron charge	C
Rp	Parallel resistance (Shunt resistance)	Ω
r_{Rec}	Recombination rate	
Rs	Serial resistance	Ω
Tc	Temperature of solar device	°C
Ts	Temperature of the sun	°C
Voc	Open circuit voltage	V
V_{th}	Thermal voltage	V
η_{pe}	Photoelectric efficiency	
η_{te}	Thermoelectric efficiency	
λ	Wavelength	nm

List of common abbreviations used throughout the text.

BHJ	Bulk hetero junction
CAA	Current active area
CB	Conduction band
CB	Chlorobenzene
CTEA	Charge transfer exciton absorption
DEA	Donor-acceptor exciton absorption
DSSC	Dye-sensitized solar cells
FBSC	Fiber based solar cell
HeOI	Heterogeneous optical intensity
Homo	Lowest unoccupied molecular orbital
HoOI	Homogeneous optical intensity
IAA	Illumination active area
ITO	Tin indium oxide
Lumo	Highest occupied molecular orbital
OCGOPV	Optical confinement geometry organic photovoltaics
OCGPV	Optical confinement geometry photovoltaics
OPV	Organic photovoltaics
P3HT	Poly-3(hexylthiophene)
PCBM	Phenyl-C61-butyric acid methyl ester
PEDOT: PSS	Poly(3,4-ethylenedioxythiophene)-poly(styrenesulfonate)
PV	Photovoltaic
VB	Valence band
ZNR	Zinc oxide nano-rod

List of Figures

List of Tables

Chapter 1
Introduction

1.1 Solar Energy and Solar Cell

As our universe continues to evolve, entropy will continue to increase. In this process, however, some acceptor systems, through absorbing power, can evolve toward a high-ordered level, at the expense of enhancing another donor-system's entropy. Fortunately, the earth is an acceptor-system getting the energy from the Sun, a donor system. The sun is exporting energy in the form of light, such that materials on earth can evolve into higher levels: from inorganic compounds to organic macromolecular compounds, then to protein, the basic unit of all living beings, ultimately benefiting human and human society. In the 0.6 billion years' evolution, solar energy has played a very important role as a driving force, and has generated most of the world's food supply, and generated countless tons of coal, oil and gas which has been heavily over-utilized by modern technologies. Without the sun, the earth would be a frozen, silent planet with nothing but inorganic components.

The advancement of recent civilization can be divided into three main stages:[1] the Working Machine Revolution (year 1764–1830), Transmission Machine Revolution (year 1830–1945) and Controlling Machine Revolution (year 1945-present), Coal was the main power source in the first stage, while gasoline and diesel oil gave way to advanced transmission machines without heavy coal-burning stoves, which were then replaced by electricity from nuclear or solar energy, which have allowed advanced machines to operate more efficiently during the present era. Nowadays, solar power has become relevant as a source of green energy and is certainly a safe power source, which has become increasingly important in the wake of the various disasters surrounding nuclear power plants over the last 30 years. Therefore, solar energy will be the main future power source for human society.

The bright future always has a long and tortuous script in history. In the last half of the nineteenth century, engineers experimented with selenium for detecting flaws of the wires they submerged under the sea for the world's first seamless communication network through transoceanic telegraph cables. Early researchers found selenium's performance was related to the sunlight, and this phenomenon aroused

Y. Li, *Three Dimensional Solar Cells Based on Optical Confinement Geometries*,
Springer Theses, DOI 10.1007/978-1-4614-5699-5_1,

the interest of many scientists, including William Grylls Adams and his student, Richard Evans Day. They found that light could cause selenium to generate current after a series of experiments. This discovery proved the theoretical possibility that photons with enough power hitting material like selenium could knock poorly-bound electrons from their orbits, which demonstrated the probability of transferring light into electricity.

The first silicon solar cell was invented by Gerald Pearson and Calvin Fuller after an accidental discovery related with their transistor research led to a breakthrough in 1953. The electric output of their solar cell was five times greater than the best selenium cell at that time. Even The New York Times proclaimed that their work "may mark the beginning of a new era, leading eventually to the realization of one of mankind's most cherished dreams: the harnessing of the limitless energy of the sun for the uses of civilization." Two years later, the American government announced its intention of launching a satellite into orbit around the earth, since this was the only feasible way to harness the full power of direct sunlight. This first satellite, named Vanguard, also marked the first application of the silicon solar cell. However, solar cells, as an industrialized product for the market, have always had one underlying problem that held back widespread adoption: price. Daryl Chapin, one of the inventors of the solar cell, wondered: "What to do with our new baby." This problem was addressed by Dr. Elliot Berman in 1968, by starting with lower grade, and therefore cheaper, silicon and finishing up with less expensive materials for packaging the modules. As a result, the price dropped to $20 per watt from $200 per watt.

After that, the application of solar modules spread to a wide variety of fields. The oil companies were the first major customers to use the solar cell as a power source of warning lights and horns in the Gulf of Mexico. Solar cells were then applied as the power source of crossing signals in southern railroads. In the late 1970s, Australia tried to develop large photovoltaic-run telecommunication networks linking together multiple towns, such that people could dial long distance directly instead of having to call the operator. Such advances have allowed the popularity of photovoltaics to grow, allowing an ever-increasing number of people to enjoy its benefits. Like wireless internet released people from offices filled with cables, solar power has the ability to release machines, instruments, and even towns from the tethered restrictions of limited regional power supply.

In the past 30 years, many governments have spent a great deal of money in developing the solar industry. In the next 50 years, it is anticipated that solar power will take a leading role in global energy production, as Fig. 1.1 shows. Depending which study one reads, energy demand is expected to at least double by 2050, and quadruple by 2100, which necessitates an additional production capacity of approximately 500EJ till 2050 [2]. In terms of Fig. 1.1, we would estimate that solar energy will supply 140 TW by 2050. This simply means that the market is very large for such technologies and is likely to embrace several different technology platforms. Nowadays, silicon solar modules have been incorporated into millions of buildings and billions of instruments. However, the power stations behind silicon solar cells, unlike those of oil and coal, still have many problems behind it,

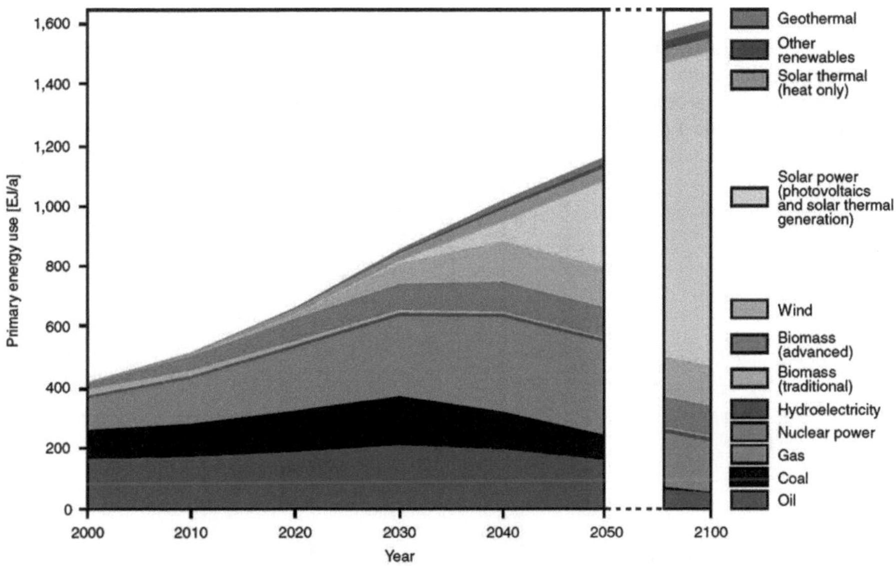

Fig. 1.1 Transforming the global energy mix: The exemplary path until 2050/2100. (Source: WBGU, World in Transition Towards Sustainable Energy Systems, at http://www.wbgu.de/) [2]

primarily because silicon is the raw material of the semiconductor industry, so that the price is hardly further reduced. Therefore, scientists all over the world have explored a variety of other materials and even solar cell types for the replacement of silicon. Close relatives to single-crystal silicon, such as polycrystalline silicon and amorphous silicon, are receiving much attention.[3] More approaches such as polycrystalline thin film $CuInGaSe_2$ [4] or CdTe [5], multijunction III-V devices [6], dye–sensitized or titania solar cell,[7] and conjugated polymer or those that are nanomaterial-enhanced [8] are also gathering much interest. However, instead of the exploration in materials, this thesis focused on a three dimensional architecture, which is to confine the light, and improve the performance of solar cell.

1.2 Physics of Photovoltaics

1.2.1 Photovoltaic Effect

Figure 1.2 shows the solar radiation spectrums for Extraterrestrial Radiation (solar spectrum at top of atmosphere) at mean Earth-Sun distance and Solar Spectral Irradiance at Air Mass 1.5 (spectral radiation from solar disk plus sky diffuse and diffuse reflected from ground on south-facing surface tilted 37° from horizontal, i.e., the surface normal points to the sun, at an elevation of 41.81° above the horizon). The solar energy is primarily distributed from 300 to 1,700 nm, and its

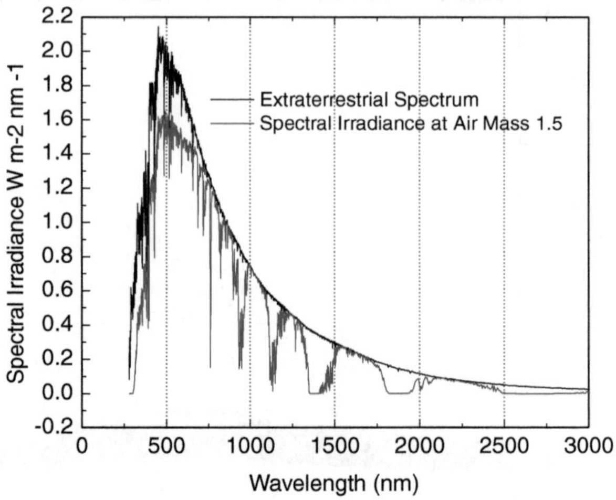

Fig. 1.2 Solar radiation spectrums for extraterrestrial radiation at mean earth-sun distance and solar spectral irradiance at air mass 1.5 [10]

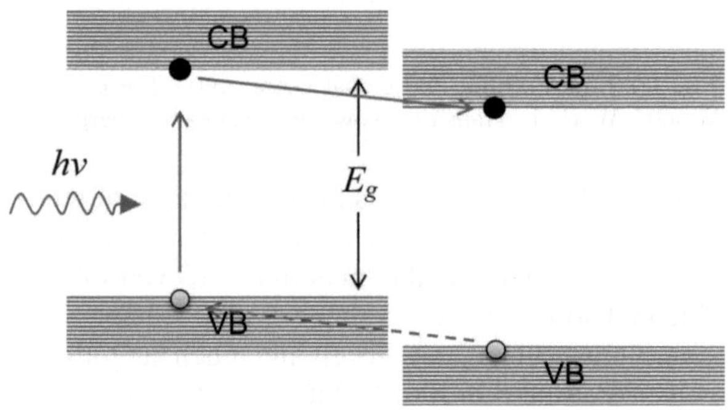

Fig. 1.3 How does active layer absorb a photon and generate a free electron

maximum peak is located at around 530 nm. The sun produces light with a distribution similar to what would be expected from a 5,525 K (5,250°C) blackbody, which is approximately the sun's surface temperature. As light passes through the atmosphere, some is absorbed by gases with specific absorption bands. Additional light is redistributed by Rayleigh scattering, which is responsible for the atmosphere's blue color [9].

When a photon with energy *hv* hits an atom, the electron in valence band (VB) is exited to the conduction band (CB) as a free electron and leave a hole at CB, as Fig. 1.3 shows.

If another material has a lower CB to accept this free electron, and this electron would not recombine with hole in the form of phonon or emission, it will contribute to the "current" in external circuit. Therefore, the maximum absorption (unit W/m^2) at certain band gap E_g is obtained by Eq. 1.1.

$$E_{\max} = \int_0^{\frac{ch}{qE_g}} S_{AM1.5} d\lambda \tag{1.1}$$

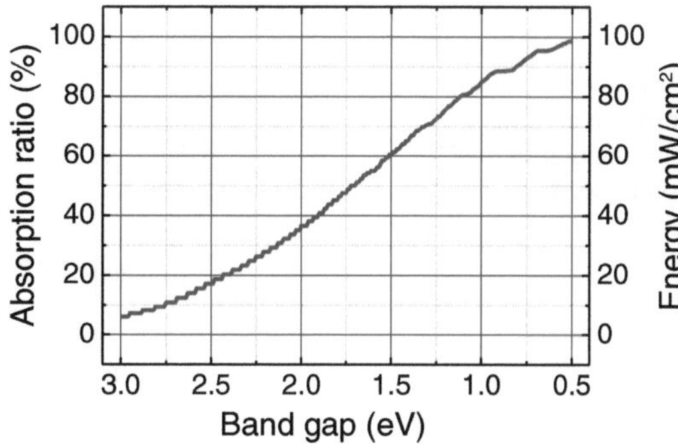

Fig. 1.4 Maximum absorption versus absorber's band gap under illumination of AM1.5 solar spectrum

Where, $S_{AM1.5}$ is Solar Spectral Irradiance at Air Mass 1.5. C, h, c and q are one coulomb (defined as the charge transported by a steady current of one ampere in 1 s, $C = 6.24 \times 10^{18}$), Planck constant, the speed of light in vacuum and electron charge; λ is the wavelength of light.

Consequently, the maximum absorption versus absorber's band gap is shown in Fig. 1.4. If not taking into account output voltage, the best material for absorption should have the band gap lower than 0.5 eV.

Although a material with a band gap of 1.1 eV (1,100 nm) is capable of absorbing 80% of the solar irradiation power at the earth's surface (for 100% absorptivity), their current conversion will lead to a huge loss due to the Photoelectric Effect, because one photon can only contribute one electron even though its energy is higher than that of band gap's. Therefore, the real current generation and corresponding energy loss is describe by Eqs. 1.2 and 1.3

$$J_{\max} = \int_0^{\frac{ch}{qE_g}} \frac{S_{AM1.5}}{Chc/\lambda} d\lambda \qquad (1.2)$$

$$E_{loss} = \int_0^{\frac{ch}{qE_g}} \left(hc/\lambda q - E_g\right) d\lambda \qquad (1.3)$$

Temporarily, neglecting the influence of an acceptor's orbital and device's defects to voltage, the open circuit voltage (V_{OC}) can be estimated as Eq. 1.4.

$$V_{oc} = E_g/q \qquad (1.4)$$

In terms of Eqs. 1.2 and 1.4, for the ideal solar cell with Fill Factor (FF) =1, the theoretical maximum efficiency of absorber is described as Eq. 1.5.

$$Eff_{\max} = V_{oc} \times J_{\max} \times FF = \frac{E_g}{q} \int_0^{\frac{ch}{qE_g}} \frac{S_{AM1.5}}{Chc/\lambda} d\lambda \qquad (1.5)$$

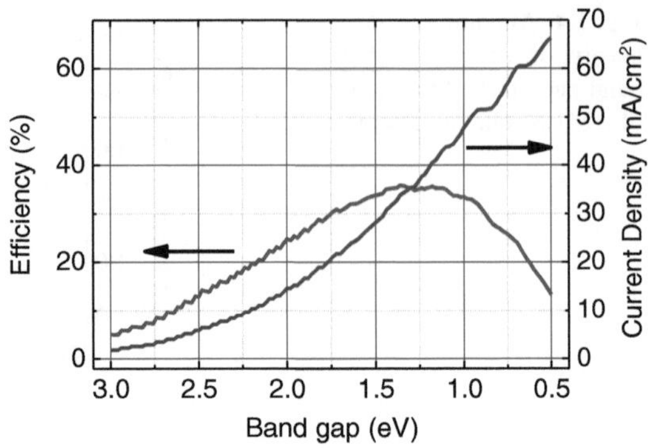

Fig. 1.5 Efficiency and current density versus absorber's band gap respectively under illumination of AM1.5 solar spectrum

The following Fig. 1.5 shows the maximum current generation and corresponding maximum efficiency determined by Eq. 1.5. The limit from photoelectric effect to absorbing material leads to a huge reduction of efficiency, in which even the maximum value at 1.3 eV band gap is only 35%. Furthermore, after considering the Shockley-Queisser limit on recombination rate, this efficiency is reduced to 31.0% for Eg = 1.3 eV for Ts = 6,000 K and Tc = 300 K, where Ts and Tc are the temperature of the sun and solar cell, respectively [11].

However, some nonlinear optical methods may break this limit, such as down-conversion. As mentioned above, the major loss in photovoltaic devices is that cells absorb high energy photons but excite electrons having the low potential equal to band gap's, and the rest of the energy is lost in the form of thermal. The down-conversion of incident photons into two or more photons with lower energies can be achieved, in principle, with every three-level system. The absorption of a high-energy photon leads to an electronic transition from the lowest level to the highest excited level. A two-step recombination of the electron to the lowest level via the intermediate level is accompanied by the emission of two lower energy photons, if both steps are radiatively efficient [12]. Therefore, if one higher photon can generate more than one electron–hole pair or exciton, the theoretical max efficiency based on this system would be enhance to a higher level. This efficiency can be estimated using Eqs. 1.6 and 1.7 by adding a function quotient () into Eqs. 1.2 and 1.5.

$$J_{\max} = \int_0^{\frac{ch}{qE_g}} Quotient(\frac{hc}{\lambda}, E_g) \frac{S_{AM1.5}}{Chc/\lambda} d\lambda \tag{1.6}$$

$$Eff_{\max} = V_{oc} \times J_{\max} \times FF = \frac{E_g}{q} \int_0^{\frac{ch}{qE_g}} Quotient(\frac{hc}{\lambda}, E_g) \frac{S_{AM1.5}}{Chc/\lambda} d\lambda \tag{1.7}$$

Comparing with the current curve in Fig. 1.5, the down-conversion effect could dramatically improve the current generation, as shown in Fig. 1.6.

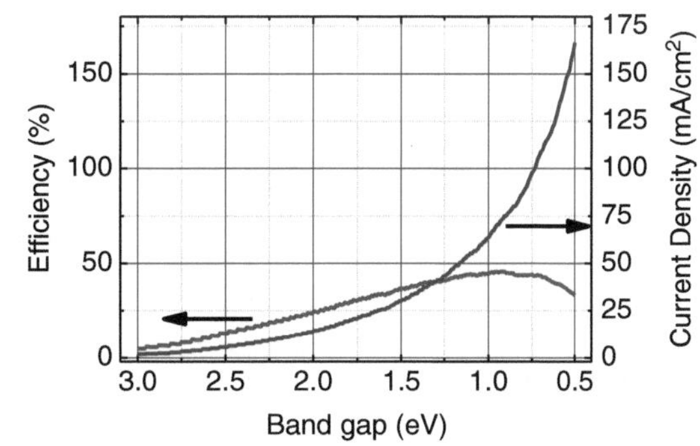

Fig. 1.6 Efficiency and current density versus absorber's band gap with under-conversion, under illumination of AM1.5 solar spectrum

For further consideration of a real photovoltaic device, there are more factors influencing open circuit voltage (V_{oc}) including temperature T, concentration n_e (n_h) of electrons (holes), and intrinsic concentration n_i of electrons and holes as shown in Eq. 1.8

$$V_{oc} = \frac{1}{q}(\eta_e + \eta_h) = \frac{k_B T}{q}\ln\left(\frac{n_e n_h}{n_i^2}\right) = \frac{E_g}{q} - \frac{k_B T}{q}\ln\left(\frac{N_c N_v}{n_e n_h}\right) \tag{1.8}$$

$$n_i^2 = N_c N_v \exp\left(-\frac{E_g}{k_B T}\right) \tag{1.9}$$

Where, N_c and N_v are effective density of states in conduction bands and valence bands, respectively. This can be connected with recombination rate r_{Rec}, thickness d and photocurrent J_{ph} in device, in Eq. 1.10 [13].

$$V_{oc} = \frac{E_g}{q} - \frac{k_B T}{q}\ln\left(\frac{q d r_{\mathrm{Rec}} N_c N_v}{J_{ph}}\right) - \Delta \tag{1.10}$$

The energy shift Δ in Eq. 1.10 represents the voltage loss due to fabrication defect. For organic solar cells, the band gap E_g is replaced by the difference between the donor's Highest Occupied Molecular Orbital (HOMO) and the acceptor's Lowest Unoccupied Molecular Orbital (LUMO) in Eq. 1.11.[14]

$$V_{oc} = \frac{1}{q}\left(E_{Lumo}^{Accepter} - E_{Homo}^{Donor}\right) - \frac{kT}{q}\ln\left(\frac{q d r_{\mathrm{Rec}} N_c N_v}{J_{ph}}\right) - \Delta \tag{1.11}$$

1.2.2 Light Distribution in Multilayers

As shown in Fig. 1.7, solar devices have many stacked layers, and most of them are of the scale of nanometer, which is the same order of magnitude as visible light. To know the light distribution in these layers with nanometer thickness, it is

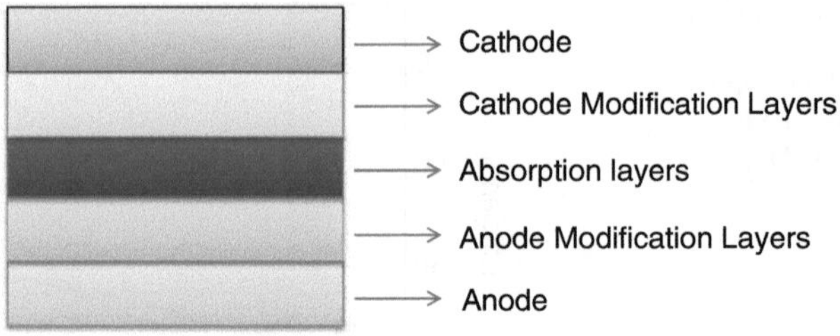

Fig. 1.7 Architecture of conventional solar cell

necessary to take into account the interference between interfaces. A transfer matrix method is introduced to calculate the optical field and absorption. The optical field distribution gives the current generation; it's active layer and steady-state current distribution determines the current output on certain applied voltage.

Figure 1.7 shows that absorption layers, cathode modification layers (CML) and anode modification layers (AML) are inserted between cathode and anode. In some cases, they have only one layer or one blend layer. For example, mostly the blended bulk-heterojunction layer (as absorption layer) is used to replace the structure of Donor layer/Acceptor layer, for 100% dissociation; Also, CML could be either single hole block layer (HBL) or the stacked layers consist of optical spacer [15–17] and EBL and other layers near cathode. (The similar situation for AML.) Therefore, there are far more than one layer influencing light transmits in devices. Usually for large scale media of millimeter's or centimeter's length that is much larger than wavelength of solar light (300 ~ 1,700 nm), The Beer–Lambert law (Eq. 1.12) is usually used to describe the light absorption and the generation of exciton [18–20]. However, in this expression, reflection at the front surface of device and the optical mode structure inside the device are neglected, i.e. this equation is applied to the film having the thickness much bigger than the wavelength of incident light.

$$I = I_0 e^{-\alpha x} \tag{1.12}$$

For these layers of nanometer thickness, the wave nature of light should be taken into account to calculate light distribution in multi-layers by a transfer matrix method [21, 22].

In Fig. 1.8, the total electric field $E_j(x)$ in an arbitrary plane in layer *j* at a distance x to the left boundary *(j-1)j* can be expressed as the sum of two components : $E_j^+$and E_j^- propagating in the positive x direction and negative x direction:

$$E_j(x) = E_j^+(x) + E_j^-(x) \tag{1.13}$$

Now, we need to find an equation to connect $E_j^+$and E_j^- with the incident light intensity and the complex refraction index $\tilde{n}_j$ which we know. The electric field at the two ends in air are represented as E_0^+ and E_0^-, and E_{m+1}^+ and E_{m+1}^-, which are related by a transfer matrix S:

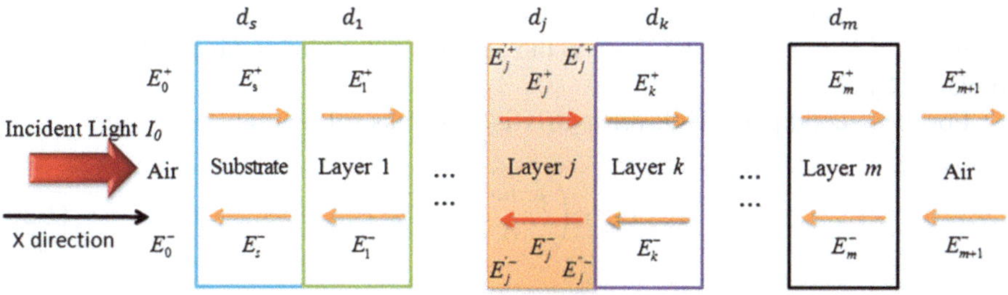

Fig. 1.8 Multilayer stack (layer i = 1, 2, ... m) are connected with substrate, and both outsides are in air. Light goes through each layer with optical interference (complex refractive index). The optical electric field has two components: E_j^+ and E_j^-, which propagate in the positive × direction and negative × direction respectively. The incident light for each wavelength is in the AM 1.5 G spectrum [10]. All variables subscripted with 0 and S represent values for air and for substrate, respectively. Layer m is mostly metallic cathode, like Al and Ag

$$\begin{pmatrix} E_0^+ \\ E_0^- \end{pmatrix} = S \begin{pmatrix} E_{m+1}^+ \\ E_{m+1}^- \end{pmatrix} \tag{1.14}$$

Where, E_0^+ is correlated with the incident light intensity $I_0(\lambda)$ (obtained from AM 1.5 G spectrum [10]) as

$$I_0(\lambda) = \frac{1}{2} c\varepsilon_0 \left| E_0^+(\lambda) \right|^2 \tag{1.15}$$

While expanding the transfer matrix S, there are two behaviors in whole light propagating process: behavior of refraction and reflection at the interface between two layers j and k, and behavior of transmission in layer j. The first part is named as interface matrix I_{jk}:

$$I_{jk} = \frac{1}{t_{jk}} \begin{pmatrix} 1 & r_{jk} \\ r_{jk} & 1 \end{pmatrix} \tag{1.16}$$

r_{jk} and t_{jk} are the Fresnel complex reflection and transmission coefficients at interface j and k. For TE wave and TM wave, the Fresnel complex reflection and transmission coefficients can be described as the function of incident angle ϕ_0

$$\text{TE wave } \mathrm{r_{jk}} = \frac{\mathrm{q_j} - \mathrm{q_k}}{\mathrm{q_j} + \mathrm{q_k}}, \ \mathrm{t_{jk}} = \frac{2\mathrm{q_j}}{\mathrm{q_j} + \mathrm{q_k}} \tag{1.17}$$

$$\text{TM wave } \mathrm{r_{jk}} = \frac{\tilde{\mathrm{n}}_{\mathrm{k}}^2 \mathrm{q_j} - \tilde{\mathrm{n}}_{\mathrm{j}}^2 \mathrm{q_k}}{\tilde{\mathrm{n}}_{\mathrm{k}}^2 \mathrm{q_j} + \tilde{\mathrm{n}}_{\mathrm{j}}^2 \mathrm{q_k}}, \ \mathrm{t_{jk}} = \frac{2\tilde{\mathrm{n}}_{\mathrm{k}} \tilde{\mathrm{n}}_{\mathrm{j}} \mathrm{q_j}}{\tilde{\mathrm{n}}_{\mathrm{k}}^2 \mathrm{q_j} + \tilde{\mathrm{n}}_{\mathrm{j}}^2 \mathrm{q_k}} \tag{1.18}$$

Where

$$q_j = \tilde{n}_j \cos\phi_j = \left[\tilde{n}_j^2 - \eta_0^2 \sin\phi_0\right]^{1/2} \tag{1.19}$$

$\tilde{n}_j = n_j + ik_j$ is the complex refraction index, which is a function of wavelength λ, namely, $\tilde{n}_j(\lambda) = n_j(\lambda) + ik_j(\lambda)$. ϕ_j is the angle of refraction in layer j, but could be replaced by ϕ_0.

For the second part, the layer matrix is shown as following:

$$L_j = \begin{pmatrix} e^{-i\xi_j d_j} & 0 \\ 0 & e^{i\xi_j d_j} \end{pmatrix} \tag{1.20}$$

Where

$$\xi_j = \frac{2\pi}{\lambda} q_j \tag{1.21}$$

d_j is the thickness of layer j. $\xi_j d_j$ is layer phase thickness. By combining the interface matrix of Eq. 1.16 and layer matrix of Eq. 1.20 into total system transfer matrix S, it can be expressed as

$$S = \begin{pmatrix} S_{11} & S_{12} \\ S_{21} & S_{22} \end{pmatrix} = \left(\prod_{v=1}^{m} I_{(v-1)v} L_v \right) I_{m(m+1)} \tag{1.22}$$

$$\begin{pmatrix} E_0^+ \\ E_0^- \end{pmatrix} = \begin{pmatrix} S_{11} & S_{12} \\ S_{21} & S_{22} \end{pmatrix} \begin{pmatrix} E_{m+1}^+ \\ E_{m+1}^- \end{pmatrix} \tag{1.23}$$

For calculating the electric field in layer j, Eq. 1.23 need be expanded as an upstream system S_j' and a downstream system S_j'', which represent the left transfer matrix and the right transfer matrix:

$$S = S_j' L_j S_j'' = \begin{pmatrix} S_{j11}' & S_{j12}' \\ S_{j21}' & S_{j22}' \end{pmatrix} L_j \begin{pmatrix} S_{j11}'' & S_{j12}'' \\ S_{j21}'' & S_{j22}'' \end{pmatrix} \tag{1.24}$$

Where

$$S_j' = \left(\prod_{v=1}^{m} I_{(v-1)v} L_v \right) I_{(j-1)j} \tag{1.25}$$

$$S_j'' = \left(\prod_{v=j+1}^{m} I_{(v-1)v} L_v \right) I_{m(m+1)} \tag{1.26}$$

As shown in Fig. 1.8, in the layer j, the electric field on both interface are expressed as $E_i'^{+}$ and $E_i'^{-}$, and $E_i''^{+}$ and $E_i''^{-}$, which related to upstream system and downstream system:

$$\begin{pmatrix} E_0^+ \\ E_0^- \end{pmatrix} = S_j' \begin{pmatrix} E_j'^+ \\ E_j'^- \end{pmatrix} = \begin{pmatrix} S_{j11}' & S_{j12}' \\ S_{j21}' & S_{j22}' \end{pmatrix} \begin{pmatrix} E_j'^+ \\ E_j'^- \end{pmatrix} = \begin{pmatrix} S_{j11}'E_j'^+ + S_{j12}'E_j'^- \\ S_{j21}'E_j'^+ + S_{j22}'E_j'^- \end{pmatrix} \tag{1.27}$$

$$\begin{aligned} \begin{pmatrix} E_j''^+ \\ E_j''^- \end{pmatrix} &= S_j' \begin{pmatrix} E_{m+1}^+ \\ E_{m+1}^- \end{pmatrix} = \begin{pmatrix} S_{j11}'' & S_{j12}'' \\ S_{j21}'' & S_{j22}'' \end{pmatrix} \begin{pmatrix} E_{m+1}^+ \\ E_{m+1}^- \end{pmatrix} \\ &= \begin{pmatrix} S_{j11}''E_{m+1}^+ + S_{j12}''E_{m+1}^- \\ S_{j21}''E_{m+1}^+ + S_{j22}''E_{m+1}^- \end{pmatrix} \end{aligned} \tag{1.28}$$

And they are connected by the layer matrix L_j:

$$\begin{pmatrix} E_j'^+ \\ E_j'^- \end{pmatrix} = L_j \begin{pmatrix} E_j''^+ \\ E_j''^- \end{pmatrix} = \begin{pmatrix} e^{-i\xi_j d_j} & 0 \\ 0 & e^{i\xi_j d_j} \end{pmatrix} \begin{pmatrix} E_j''^+ \\ E_j''^- \end{pmatrix} = \begin{pmatrix} e^{-i\xi_j d_j} E_j''^+ \\ e^{i\xi_j d_j} E_j''^- \end{pmatrix} \tag{1.29}$$

Also, it indicates the two components of $E_j(x)$ in an arbitrary plane in layer j at a distance x to the left boundary $(j\text{-}1)j$ can be given by:

$$\begin{pmatrix} e^{i\xi_j x} & 0 \\ 0 & e^{-i\xi_j x} \end{pmatrix} \begin{pmatrix} E_j'^+ \\ E_j'^- \end{pmatrix} = \begin{pmatrix} E_j^+ \\ E_j^- \end{pmatrix} = \begin{pmatrix} e^{-i\xi_j (d_j - x)} & 0 \\ 0 & e^{i\xi_j (d_j - x)} \end{pmatrix} \begin{pmatrix} E_j''^+ \\ E_j''^- \end{pmatrix} \tag{1.30}$$

$$E_j(x) = e^{-i\xi_j x} E_j''^+ + e^{i\xi_j x} E_j''^- \tag{1.31}$$

Mostly, the layer m is metallic layer, so we can assume $E_{m+1}^- = 0$. Thus, we can rewrite Eqs. 1.27, 1.28 and 1.29 step by step

$$\begin{pmatrix} E_j''^+ \\ E_j''^- \end{pmatrix} = \begin{pmatrix} S_{j11}''E_{m+1}^+ + S_{j12}''E_{m+1}^- \\ S_{j21}''E_{m+1}^+ + S_{j22}''E_{m+1}^- \end{pmatrix} = \begin{pmatrix} S_{j11}''E_{m+1}^+ \\ S_{j21}''E_{m+1}^+ \end{pmatrix} \tag{1.32}$$

$$\begin{pmatrix} E_j'^+ \\ E_j'^- \end{pmatrix} = \begin{pmatrix} e^{-i\xi_j d_j} E_j''^+ \\ e^{i\xi_j d_j} E_j''^- \end{pmatrix} = \begin{pmatrix} e^{-i\xi_j d_j} S_{j11}'' E_{m+1}^+ \\ e^{i\xi_j d_j} S_{j21}'' E_{m+1}^+ \end{pmatrix} \tag{1.33}$$

$$\begin{aligned} \begin{pmatrix} E_0^+ \\ E_0^- \end{pmatrix} &= \begin{pmatrix} S_{j11}'E_j'^+ + S_{j12}'E_j'^- \\ S_{j21}'E_j'^+ + S_{j22}'E_j'^- \end{pmatrix} \\ &= \begin{pmatrix} S_{j11}' e^{-i\xi_j d_j} S_{j11}'' E_{m+1}^+ + S_{j12}' e^{i\xi_j d_j} S_{j21}'' E_{m+1}^+ \\ S_{j21}' e^{-i\xi_j d_j} S_{j11}'' E_{m+1}^+ + S_{j22}' e^{i\xi_j d_j} S_{j21}'' E_{m+1}^+ \end{pmatrix} \end{aligned} \tag{1.34}$$

We obtained the relationship of E_{m+1}^+ and E_0^+

$$E_{m+1}^{+} = \frac{E_0^{+}}{(S'_{j11}e^{-i\xi_j d_j}S''_{j11} + S'_{j12}e^{i\xi_j d_j}S''_{j21})} \tag{1.35}$$

Putting Eqs. 1.35 into 1.31 and 1.32, the total electric field $E_j(x)$ finally can be expressed as

$$E_j(x) = E_j^{+}(x) + E_j^{-}(x) = \frac{\left(S''_{j11}e^{-i\xi_j(d_j-x)} + S''_{j21}e^{i\xi_j(d_j-x)}\right)}{\left(S'_{j11}S''_{j11}e^{-i\xi_j d_j} + S'_{j12}S''_{j21}e^{i\xi_j d_j}\right)}E_0^{+} \tag{1.36}$$

Further this can be used to relate the total electric field with absorption and then to obtain current density. Based on the pointing theorem,[23] that can be interpreted as a statement of conservation of energy Q(x), which is the (time average) energy flow dissipation per time unit at the point x.

$$Q(x) = \langle -\nabla \cdot \mathbf{S} \rangle \tag{1.37}$$

Where

$$\mathbf{S} = E \times H \tag{1.38}$$

Because time averaging of a product of $\tilde{A} = A_0 e^{i(kx-wt)}$ and $\tilde{B} = B_0 e^{i(kx-wt)}$ can be performed as $\frac{1}{2}Re(\tilde{A}\tilde{B}^*)$, and

$$-\nabla \cdot \mathbf{S} = \left(H\frac{\partial \mathbf{B}}{\partial t} + E\frac{\partial \mathbf{D}}{\partial t}\right) \tag{1.39}$$

Using Maxwell equations, one can show that

$$Q(x) = \frac{1}{2}c\varepsilon_0 \alpha_j n_j |\mathbf{E}(x)|^2 \tag{1.40}$$

Where

$$\alpha_j = \frac{4\pi k_j}{\lambda} \tag{1.41}$$

α_j is the absorption coefficient. Then, put Eqs. 1.15 and 1.36 into 1.40

$$Q(x,\lambda) = \alpha_j n_j I_0(\lambda)\left|\frac{\left(S''_{j11}e^{-i\xi_j(d_j-x)} + S''_{j21}e^{i\xi_j(d_j-x)}\right)}{\left(S'_{j11}S''_{j11}e^{-i\xi_j d_j} + S'_{j12}S''_{j21}e^{i\xi_j d_j}\right)}\right|^2 \tag{1.42}$$

The transfer matrix method is applied to thin films with thickness near visible light wavelength. However, since the large thickness of substrate is of order ~mm, it cannot be included directly in the transfer matrix calculation. By using Eqs. 1.17 ~ 1.19, and 1.23, the reflectance and transmittance of substrate and stacked multi-layers can be expressed as:

$$R_S = |r_{0S}|^2 = \left|\frac{q_0 - q_S}{q_0 + q_S}\right|^2 \tag{1.43}$$

$$T_S = |t_{0S}|^2 = \left|\frac{2q_S}{q_0 + q_S}\right|^2 \tag{1.44}$$

$$R_m = \left|\frac{S_{21}}{S_{11}}\right|^2 \tag{1.45}$$

$$T_m = \left|\frac{1}{S_{11}}\right|^2 \tag{1.46}$$

Due to light oscillation in substrate, the total transmittance of whole structure is

$$T_{total} = T_S e^{-\alpha_S d_S} \left(\sum_{i=0}^{\infty} \left(R_S R_m e^{-\alpha_S d_S}\right)^i \right) T = \frac{T_S T e^{-\alpha_S d_S}}{1 - R_S R_m e^{-\alpha_S d_S}} \tag{1.47}$$

Where, $\alpha_S = 4\pi k_S/\lambda$ is the absorption coefficient in substrate.

Thus, the intensity of light on multilayer from substrate is given by

$$\left|E_S^{\prime\prime +}\right|^2 = \frac{T_S e^{-\alpha_S d_S}}{1 - R_S R_m e^{-\alpha_S d_S}} \left|E_0^+\right|^2 \tag{1.48}$$

And in terms of Eq. 1.15

$$I_S = \frac{T_S e^{-\alpha_S d_S}}{n_S\left(1 - R_S R_m e^{-\alpha_S d_S}\right)} I_0 \tag{1.49}$$

Then, use Eq. 1.49 to replace I_0 in Eq. 1.42

$$Q(x,\lambda) = \alpha_j n_j \frac{T_S e^{-\alpha_S d_S}}{n_S\left(1 - R_S R_m e^{-\alpha_S d_S}\right)} I_0 \left| \frac{\left(S_{j11}^{\prime\prime} e^{-i\xi_j(d_j - x)} + S_{j21}^{\prime\prime} e^{i\xi_j(d_j - x)}\right)}{\left(S_{j11}^{\prime} S_{j11}^{\prime\prime} e^{-i\xi_j d_j} + S_{j12}^{\prime} S_{j21}^{\prime\prime} e^{i\xi_j d_j}\right)} \right|^2 \tag{1.50}$$

Similarly, the electric field there can be written as

$$E_S^{''+} = \frac{t_S e^{-\xi_S d_S}}{1 - r_S r_m e^{-\xi_S d_S}} E_0^+ \tag{1.51}$$

And

$$E_j(x) = \frac{\left(S''_{j11} e^{-i\xi_j(d_j - x)} + S''_{j21} e^{i\xi_j(d_j - x)}\right)}{\left(S'_{j11} S''_{j11} e^{-i\xi_j d_j} + S'_{j12} S''_{j21} e^{i\xi_j d_j}\right)} \frac{t_S e^{-\xi_S d_S}}{1 - r_S r_m e^{-\xi_S d_S}} E_0^+ \tag{1.52}$$

Since solar light includes TE wave and TM wave and they are equal to half of solar energy, the total light energy absorption is written as Eq. 1.53 in terms of Eq.1.50.

$$Q_{totoal} = (Q_{TE}(x, \lambda) + Q_{TM}(x, \lambda))/2 \tag{1.53}$$

Now, electric field and energy absorption in each layer can be calculated by this method.

1.2.3 Drift, Diffusion and Recombination of Charge Carrier

The transfer matrix method above provides a method to calculate the optical field distribution, light absorption and corresponding current generation in active layers at certain positions. To further calculate the charge carrier behaviors, this section is to introduce how to obtain the steady-state current distribution, and find the current output on certain applied voltage.

First, to obtain the steady state current in a device, the Poisson's equation (Eq. 1.54), the continuity equations (Eqs. 1.55 and 1.56) and the drift diffusion equations (Eqs. 1.57and 1.58) are solved to find the stable solution, by using a numerical method [24].

$$\frac{\partial^2 \Psi(x)}{\partial x^2} = \frac{q}{\varepsilon}[n(x) - p(x)] \tag{1.54}$$

$$\frac{\partial n}{\partial t} - \frac{1}{q}\frac{\partial J_n(x)}{\partial x} = U(x) \tag{1.55}$$

$$\frac{\partial p}{\partial t} + \frac{1}{q}\frac{\partial J_p(x)}{\partial x} = U(x) \tag{1.56}$$

$$J_n(x) = -qn(x)\mu_n \frac{\partial}{\partial x}\Psi(x) + qD_n \frac{\partial}{\partial x} n(x) \tag{1.57}$$

$$J_p(x) = -qn(x)\mu_n \frac{\partial}{\partial x}\Psi(x) - qD_p \frac{\partial}{\partial x}p(x) \tag{1.58}$$

Where, q is the elementary charge and ε is the dielectric constant, relating the potential $\Psi(x)$ to the electron and hole densities n(x) and p(x), respectively. $J_{n(p)}(x)$ is the electron (hole) current density and $D_{n,p} = \mu_{n,p}V_t$ are the carrier of diffusion coefficients, which are assumed to obey the Einstein relation [25]. $V_t = k_BT/q$ is the thermal voltage. U(x) is the net generation rate, i.e. the difference between generation rat X(x) and recombination rate P(x) for free carriers, which is shown in Eq. 1.59.

$$U(x) = PX - (1 - P)X \tag{1.59}$$

The exciton generation rate X can be written by dividing energy dissipated per unit volume Q (Eq. 1.53) by a single photon of wavelength λ and then integrating over all wavelengths in the spectrum, in Eq. 1.60.

$$X(x) = \eta \int_{300}^{\frac{ch}{eE_g}} \frac{\lambda}{hc} Q(x,\lambda)d\lambda \tag{1.60}$$

The overall exciton dissociation probability can be written as an integral over all charge-separation distances:

$$P(E,T) = \frac{4}{a^3\sqrt{\pi}} \int_0^{\infty} p(x,E,T)x^2 e^{-(x/a)^2} dx \tag{1.61}$$

Where, x is the distance between the bound charge of exciton, E is electric field, T is the temperature, k_r is the rate at which excitons relax to the ground state, and k_d is the dissociation rate. The probability of exciton dissociation p(x,F,T) is expressed as Eq. 1.62 [22, 26].

$$p(x,E,T) = \frac{k_d(x,E,T)}{k_d(x,E,T) + k_r} \tag{1.62}$$

$$k_d(x,E,T) = \frac{3\gamma}{4\pi x^3} \exp\left(\frac{-U_b}{k_BT}\right) \frac{J_1(2\sqrt{-2b})}{\sqrt{-2b}} \tag{1.63}$$

Where, γ is the Langevin bimolecular recombination rate constant, $U_b = q^2/(4\pi\varepsilon_r\varepsilon_0 r)$ is the exciton binding energy, J_1 is the first order Bessel function, and the field parameter $b = q^3F/(8\pi\varepsilon_r k_B{}^2T^2)$. ε_r and ε_0 is material's dielectric constant and permittivity of free space, respectively.

The boundary conditions are necessary to obtain a unique solution of the system formed by Eqs. 1.54 ~ 1.58. The top contact, x = 0, is assumed to line up with the

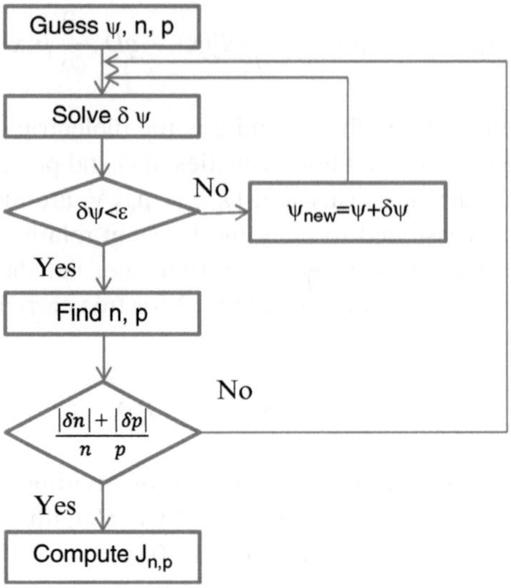

Fig. 1.9 Flow diagram of the simulation program to find the steady state current

conduction band of semiconductor and is therefore Ohmic. Also using Boltzmann statistics

$$n(0) = N_c \tag{1.64}$$

$$p(0) = N_c \exp\left(-\frac{E_{gap}}{V_t}\right) \tag{1.65}$$

Where N_c is the effective density of states of both the conduction and valence band.

Similarly, the bottom contact, x = L is assumed to be hole Ohmic:

$$n(L) = N_c\left(-\frac{E_{gap}}{V_t}\right) \tag{1.66}$$

$$p(L) = N_c \tag{1.67}$$

Since the Ohmic nature of both contacts is evidenced by current–voltage measurements on both materials, the boundary condition for the potential is written as [24]

$$\Psi(L) - \Psi(0) = E_{gap} - V_a \tag{1.68}$$

Koster introduced a scheme to solve the Poisson and continuity equations based on the work of Gummel [27] as shown in Fig. 1.9. Firstly, a guess is made for potential and carrier densities. A correction $\delta\Psi$ is found to rebuild the potential $\Psi_{new} = \Psi + \delta\Psi$, and then this process is repeated until convergence $\delta\Psi < \varepsilon$ is reached. Next, the carrier densities are calculated from the new potential by solving the continuity equations. This entire procedure is repeated until the change in the carrier densities becomes very small, namely reaching the second convergence condition.

1.2.4 Equivalent Circuit

1.2.4.1 Models of Equivalent Circuit

Single Cell

To comprehensively understand the device physics of solar cell, this section will introduce the equivalent circuit of the solar cell and calculation of the involving parameters in it, as shown in Fig. 1.10.

In Fig. 1.10, the equivalent circuit of a planar cell is described by a current source, a diode, and two resistances in series and parallel, respectively. The resistance R_p, in parallel with diodes, represents the shunts which can occur in real solar cells cross the surfaces, at pin holes in the p-n junction or at grain boundaries. The series resistance R_s accounts for all voltage drops across the transport resistances of the solar cell and its connections to a load [28]. For a conventional solar cell, the shunt resistor R_p is due to recombination of charge carriers near the dissociation site (e.g. donor/acceptor interface) and it may also include recombination farther away from the dissociation site (e.g. near electrode). The series resistor R_s considers conductivity i.e. mobility of a specific charge carrier in the respective transport medium, where the mobility is affected by space charges and traps or other barriers (hopping) [29].

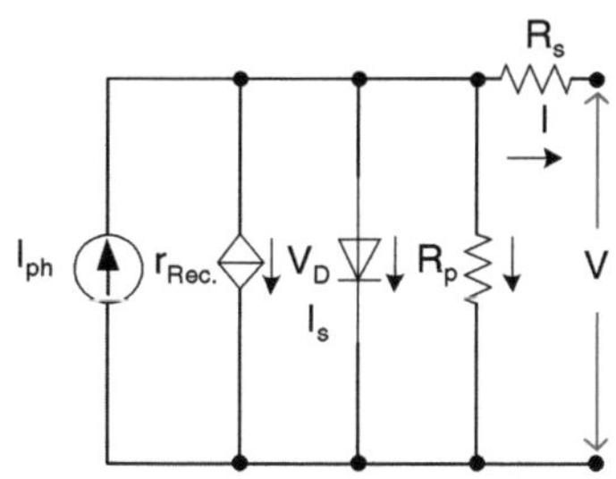

Fig. 1.10 Equivalent circuit of conventional planar solar cell. I_{ph}, $r_{Rec.}$, V_D, I_s, R_s, and R_p are photocurrent source, Recombination rate, diode voltage, reverse saturation current, series resistances and parallel resistances, respectively. I and V are the output current density and voltage of cell

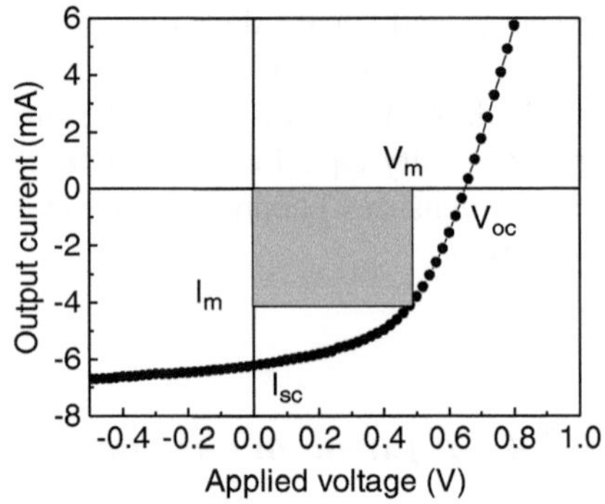

Fig. 1.11 An example of current–voltage curve of solar cell (from the device of architecture ITO/PEDOT/P3HT:PCBM/Al)

Its J-V characteristics could be expressed by the generalized Shockley equation [30]

$$I = I_{ph} - I_s\left[\exp\left(\frac{q(V - IR_s)}{n_D k_B T}\right) - 1\right] - \frac{V + R_s I}{R_p} - r_{\text{Rec.}} I_{ph} \tag{1.69}$$

Where, n_D is the diode ideality factor, here, we assume $n_D = 1$. And I_s is the reverse saturation current [31, 32]. Although there are lots of conditions influencing the recombination rate $r_{\text{Rec.}}$, the main factor is thickness while in room temperature and the same fabrication proceed.

Figure 1.11 shows an example of I-V characteristic in terms of Eq. 1.69, which is based on the architecture ITO/PEDOT/P3HT:PCBM/Al. This curve could be obtained by a numerical method: Given R_s, R_p, n_D, r_{rec}, I_s and T, the output current I are generated by solving Eq. 1.69 at each applied voltage V. In addition, the Lambert W function can also be used to obtain the list of I-V, which is discussed in Sect. 1.2.4.2.3 in details.

After normalizing the current by active area, the open circuit voltage (Voc) and short circuit current density (Jsc) are the intercept of x axis and y axis in current density-voltage (J-V array) respectively. Fill factor (FF) and efficiency are determined by the following method

$$FF = \frac{V_m J_m}{V_{oc} J_{sc}} \tag{1.70}$$

$$Eff = FF \cdot V_{oc} \cdot J_{sc} \tag{1.71}$$

Where, V_m and J_m are of the coordinates with the maximum $V_m \times J_m$ in *J-V* curve in the range of between $V_m \in [0, Voc]$, and $J_m \in [-Jsc, 0]$.

Tandem Cell

The equivalent circuit of tandem architectures can be obtained in terms of the general form shown above. Tandem cells have serial connection and parallel connection which are shown in Fig. 1.12 (a.1) and (b.1) respectively. Their simplified equivalent

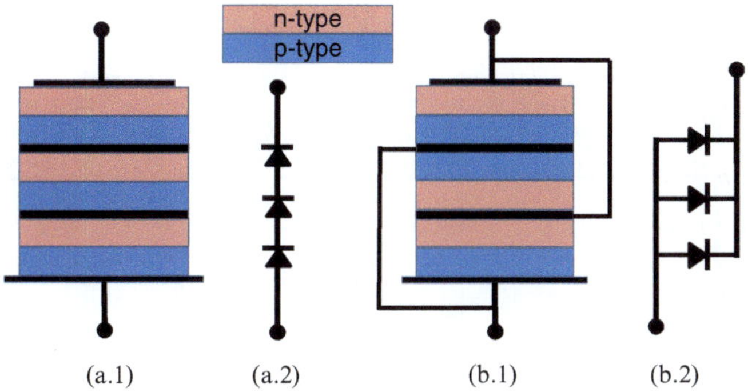

Fig. 1.12 Architectures of tandem cell and corresponding simplified equivalent circuits for serial connection (a.1 ~ a.2) and parallel connection (b.1 ~ b.2)

circuits are drawn next to them. In most tandem cell research, to calculate Isc and Voc using the simplified equivalent circuits, i.e. for serial (parallel) connection, output Jsc (Voc) is the minimum of each subcells' I_i (V_i) and the output Voc (Isc) is the sum of all subcells' I_i (V_i). However, this neglects the influences from other parameters. Taking serial connection as an example, the equivalent circuit closing real situation is shown in Fig. 1.13.

Therefore, using that of the single cell's as shown in Fig. 1.10. The voltage on each diode is generated from the parallel resistance, which equals to $I_{Ri}R_{pi}$. Therefore, the current generated by diode is described as Eq. 1.72.

$$I_{Di} = I_{0i}\left[\exp\left(\frac{qR_{pi}I_{Ri}}{n_i k_B T}\right) - 1\right] \tag{1.72}$$

In terms of Kirchhoff's principle, the current in each node should be the same, i.e. $I = I_i$. When $V = 0$, the J-V characteristics expressed by the generalized Shockley equation [33] for each cell can be written as

$$\frac{R_{pi}}{R_{si} + R_{pi}}\left\{I_{pi} - I_{0i}\left[\exp\left(-\frac{qR_p I_{R1}}{n_i k_B T}\right) - 1\right] - \frac{V_i}{R_{pi}}\right\} = I_{sc} \tag{1.73}$$

$$\sum_i^m V_i + \sum_i^m R_s I_{sc} = 0 \tag{1.74}$$

Therefore, the output short current is obtained by solving this equation set. Several parameters are necessary to this model. R_{si}, R_{pi}, I_{0i} and n_i are extracted from the IV curve of *subcell* i with the same active material and thickness. Sometimes there are two or more than two diodes to represent circuit, i.e. another term similar to the first term in the right side of Eq. 1.69 would be added, but the arithmetic method to solve this are the same as that of single diode.

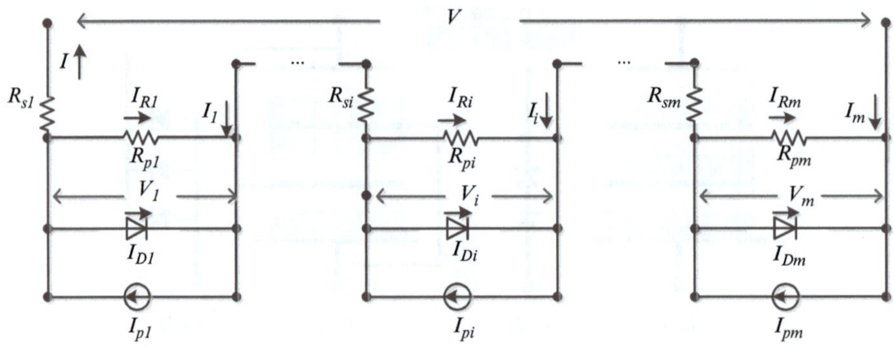

Fig. 1.13 The equivalent circuits of series tandem cell with m subcells. The subscript i is the number for each *subcell*; I_i and V_i are the output current and voltage of each *subcell* respectively; I and V are the output current and voltage of whole device respectively; n_D is the diode ideality factor; I_{0i} is the reverse saturation current; R_{si}, R_{pi}, I_{pi} the series resistance, parallel resistance, and the current passing through the parallel resistance respectively

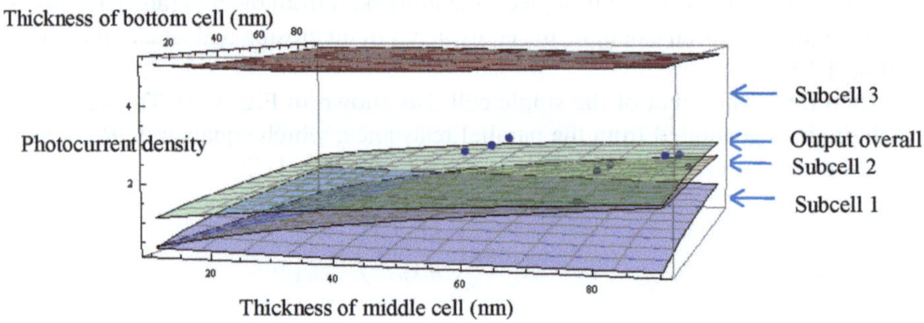

Fig. 1.14 Open current density versus the thickness of two top active layers. Surfaces are the photocurrent density generated from each subcell and whole device. *Blue points* are the experimental data

To prove this prediction, a serial tandem cell was fabricated with the three subcells, having the architecture ITO/PEDOT:PSS (20 nm)/P3HT:PCBM/Al (1 nm)/MoO_3 (15 nm)/P3HT:PCBM/Al (1 nm)/MoO_3 (15 nm)/P3HT:PCBM/Al (100 nm). As shown in Fig. 1.14, the open current generated by *subcells* are sorted as the *subcell* 3, *subcell2*, and *subcell* 1 decreasingly. It because the *subcell* 3 is the top cell absorbing light firstly and has transferred most of energy. The photocurrent of device is a value between that of the top cell and bottom cell as proved before. All experimental data are located around the prediction of photocurrent of whole device. Therefore it is proved that the equivalent circuit in Fig. 1.13 and corresponding solutions from Eqs. 1.73 and 1.74 can more accurately predict the performance of serial tandem cells.

Likewise, the equivalent circuit of parallel connection is represented as Fig. 1.15

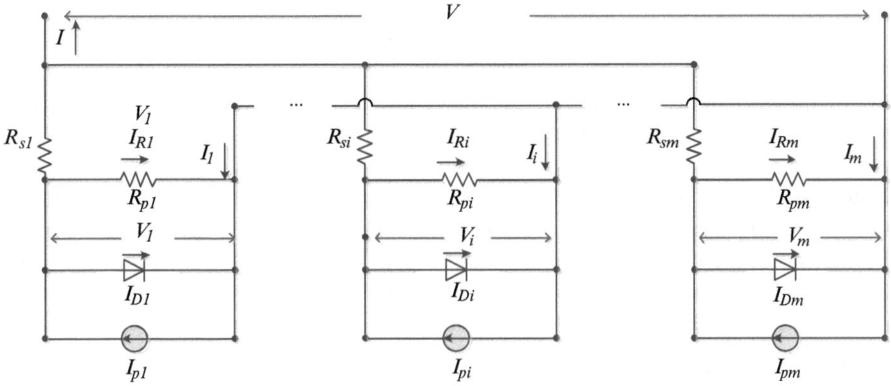

Fig. 1.15 The equivalent circuits of parallel tandem cell with m subcells. The symbols and subscripts keep the same meaning with those in Fig. 1.13

In this circuit, the current generated by diode I_{Di} is described by the same equation with Eq. 1.72. The voltage in each node should be the same, i.e. $V = V_i$. When $I = 0$, the I-V characteristics are expressed as

$$\frac{R_{pi}}{R_{si}+R_{pi}}\left\{I_{0i}\left[\exp\left(\frac{qV_i}{n_i k_B T}\right)-1\right]+\frac{V_i}{R_{pi}}-I_{pi}\right\}=0 \tag{1.75}$$

$$\sum_i^m\left(I_{pi}-I_{0i}\left[\exp\left(\frac{qV_i}{n_i k_B T}\right)-1\right]\right)R_{pi}=V_{oc} \tag{1.76}$$

By solving Eqs. 1.75 and 1.76, all parameters of subcells are obtained to predict the performance of parallel tandem cell.

The equivalent circuit of optical confinement geometry solar cell is discussed in Chap. 2.

1.2.4.2 Extracting Parameters

As discussed above, the equivalent circuit and corresponding equations are used to evaluate or predict the performance of solar device. However, there are still several unknown parameters waiting for acquisition prior to use, such as serial resistance Rs, parallel resistance Rp, ideal factor n_D, and saturation current I_s. Usually, they are extracted from current–voltage (I-V) curve. Figure 1.11 is an example of I-V curve of the architecture ITO/PEDOT/P3HT:PCBM/Al. Several extraction methods are introduced below to discuss their feature and problems.

Reciprocals of Slope at Two Point

The fastest method available to estimate series resistance and parallel resistance is the reciprocals of slope of the output curve under dark conditions at V = 0 V and V = 2 V to find the shunt and series resistances respectively as shown in Eqs. 1.77 and 1.78 [34, 35]. (Sometimes, we use two times of V_{oc} as the shunt resistance point instead of 2 V.) It is based on the assumption that the series resistance is small and the shunt resistance is large, i.e. the solar cell has reasonably good properties. Furthermore, the calculation for Rs is only valid if a 'good' voltage point is chosen for the slope calculation. In the study of organic solar cells, especially with new materials and varying solar cell architectures, the assumptions of low Rs and high Rp are not strictly valid [37].

$$R_p = -\frac{dV}{dI}\bigg|_{V=0}. \tag{1.77}$$

$$R_s = -\frac{dV}{dI}\bigg|_{V=2V\text{olt or }2\text{V}_{\text{oc}}}. \tag{1.78}$$

An Extensively Valid and Stable Method

This method has a stable solution and a rapid convergence property. It connects R_s and n_D with all other parameters, and Rs and n_D are found by fitting the data in terms of Eq. 1.79 [36]. However, since the Eq. 1.79 includes an item dV/dI, the current in data cannot be the same at any two neighboring points, such that this method does not suit for some IV with very high FF. In addition, some bad data with high s curve (namely low FF) is hard to be convergent to find a stable solution.

$$-\frac{dV}{dI} = \frac{n_D k_B T/q}{I_{sc} - I - \{V - R_{s0}(I_{sc} - I) - n_D k_B T/q\}/R_p} + R_s \tag{1.79}$$

Lambert W-function

Another simply manner to extract these parameters is through the use of a nonlinear least squared error fit where Eq. 1.69 [29, 37]. Unfortunately, since Eq. 1.69 is a transcendental equation, it is very difficult to find the numerical solutions. In order to circumvent this problem, it is convenient to use the Lambert W function described in Refs. [29, 37]. Eq. 1.69 is then written as below

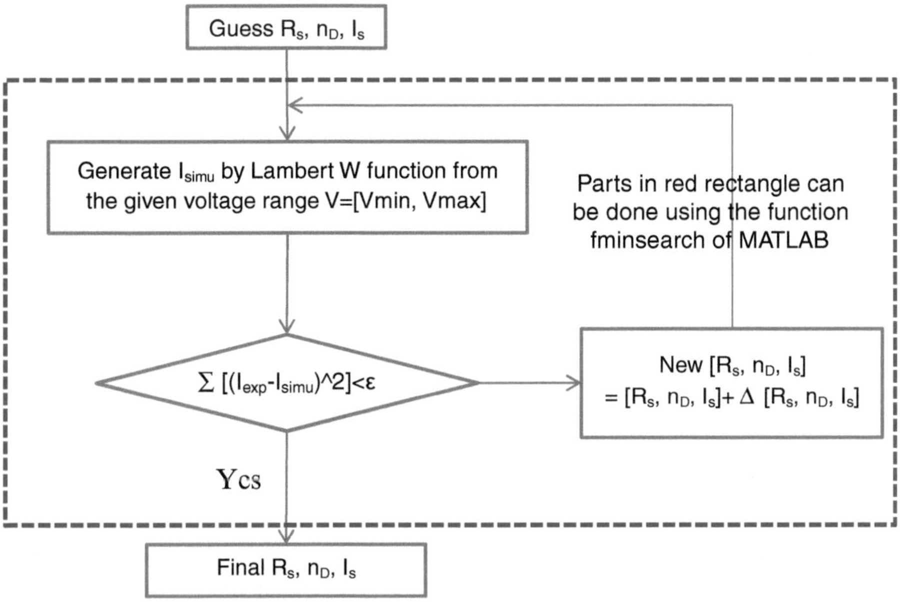

Fig. 1.16 Flow diagram of the simulation program

$$I = -\frac{V}{R_s + R_p} - \frac{n_D V_{th}}{R_s} \cdot LambertW \left[\frac{R_s I_s R_p \left[\frac{R_p(V + R_s I_s + R_s I_{ph})}{n_D V_{th}(R_s + R_p)} \right]}{n_D V_{th}(R_s + R_p)} \right] + \frac{R_p(I_s + I_{ph})}{R_s + R_{sh}} \tag{1.80}$$

Where, Lambert W function assumes I_{ph} is approximately equal to short circuit current I_{sc}; $V_{th} = k_B T/q$ is the thermal voltage. Based on Eq. 1.80, A simple and accurate (without approximations) method [37] using Lambert W function is introduced by Amit Jain. It still assumes Rp as the slope in Eq. 1.77 The unknown parameter Rs, n_D and I_s are found by a iterative searching method as shown below Fig. 1.16.

Unfortunately, this method is computationally intensive and susceptible to divergence problems as well as local minima convergence issues. Furthermore, the requirement of initial guesses infers that the researcher has some knowledge regarding the device characteristics prior to analysis, which may not always be true. [38]

Bouzidi-Chegaar Method

Bouzidi and Chegaar reported a fast method for collecting the non-exponential current terms. It can suit for some bad data with electrical noise or random errors.

The Eq. 1.69 was rewritten with the voltage as the dependent term and the current as the independent term in Eq. 1.81 [39]

$$I = \frac{n_D}{\beta} \ln \frac{I_{pA}}{I_s} + \frac{n_D}{\beta} \ln\left(1 - \frac{I_c}{I_{pA}}\right) - R_s I \tag{1.81}$$

Where,

$$\begin{cases} I_{pA} = \dfrac{I_{ph}}{1 + R_s/R_p} \\ I_0 = \dfrac{I_s}{1 + R_s/R_p} \\ \beta = \dfrac{q}{kT} \\ I_c = I_{PA} - G_A V \end{cases} \tag{1.82}$$

The Eq. 1.81 can be expressed in the common below, as it is simple to perform a simple least squares method to determine the relevant solar cell parameters (details see Ref. [39]).

$$f(I) = C_0 + C_1 + C_2 \ln\left(1 - \frac{I_c}{I_{pA}}\right) \tag{1.83}$$

The constants C_0, C_1, and C_2 are connected with the unknown parameters

$$\begin{cases} R_s = -C_1 \\ R_p = (1 - G_A R_s)/G_A \\ n = \beta C_2 \\ I_{ph} = (1 - G_A R_s)/I_{pA} \\ I_0 = I_{pA} \exp(-C_0/C_2) \\ I_s = (1 - G_A R_s)/I_0 \end{cases} \tag{1.84}$$

Where, $G_A = 1/(R_p(1 + R_s/R_p))$.

I am trying to make a web-based photovoltaic analysis platform (OPVAP, www.opvap.com) on which people can study the experimental data of solar cell and extract parameters of I-V characteristics. More numerous parameter extraction methods [40–42] will be introduced and evaluated on this platform.

Before the end of Sect. 1.2.1, there are also several classic books introducing the physics of solar cell for new Ph.D students, some of which are listed below:

- Peter Wurfel, Physics of Solar Cells, from principles to new concepts, Wiley-VCH Verlag GmbH & Co. KGaA

- Martin A. Green, Third Generation Photovoltaics advanced solar energy conversion, Springer
- Sam-Shajing Sun and Niyazi Serdar Sariciftci, Organic Photovoltaics: Mechanisms, Materials, and Devices, CRC Press

1.3 Development of Optical Confinement Geometry Solar Cell

A very high efficiency solar cell (VHESC) demonstrated 42% external conversion efficiency reported by the University of Delaware in 2008, highlighting an approach of spectral splitting and separated absorption/conversion functions to achieve solar conversion close to the Queisser limit [43] for linear light/matter interactions [11]. We propose that it is possible to make substantial further advancements in a direction parallel to that suggested by VHESC to achieve better lateral compactness, economy of fabrication and material, and increased ultimate conversion efficiency using nonlinear optical conversion, light-confinement, and programmed linear disposition of tuned absorbers in a photovoltaic geometry of coated optical fiber cavities first studied and patented by our group [44]. Originally developed as a 3D light-gathering and distribution platform for organic photovoltaics, the fiber photovoltaic geometry has already been shown to result in greater photon capture and a broader angular response than planar devices [44–46].

The key feature of VHESC devices is the use of 3D architectures to enhance light capture. There are a number of architectures that can do this, including optical fibers, aligned nanowires, V-grooves, and metamaterial lenses. In general, there are two classes of these 3D photovoltaic structures that are widely studied. The first is based on aligning nanomaterials–nanowires or nanocoaxial structures–within the absorbing medium, or with the absorbing medium around them. The key phrase here is "nano" or subwavelength. "Optical confinement" in such structures is proposed to come from scattering between the arranged objects in a loosely defined "photon localization" [47]. This could be termed a near-field approach and there are many promising examples: CdSe/Te [47] and Ge/Si [48], as well as carbon nanotubes used with Si [49] and many using organics. The confinement of the optical energy is achieved by partitioning it into the modes of an optical cavity. The solar cells based on this confinement geometry are called Optical Confinement Geometry Photovoltaics (OCGPV). In addition, this far-field approach has typically been associated with the use of high performance solid state absorbers [50].

This thesis will mainly focus on the use of the far-field approach in which light is confined in an optical cavity onto which the photoactive absorber has been added, and will also discuss the application of ZnO nano-fiber in solar devices as an example of the near-field approach. Curiously, Optical confinement geometry solar devices also have an advantageous spectral response, incident light angle dependence, better working range, and collecting more light over the day than a flat device [45, 51] making them different, in this respect, from VHESC.

Fabrication/deposition of multiple absorber layers onto 3D architectures is challenging for materials such as crystalline Si. The curvature of the substrate together with the use of polycrystalline transparent conductors (TCOs) makes growth of well-defined single crystal structures nearly impossible. For materials such as polymers or amorphous/nanocrystalline inorganics, the technical difficulties are considerably less, but the performance typically is lower (if planar devices can be used as a guide). Earlier work by our group and others [52] has shown that such geometries present extraordinarily long path lengths for the light, even in the cases of ultra-thin absorbers. Thus, we may expect this geometry to perform optimally for absorbers substantially thinner than found in planar devices (for a-Si:H this is of the order 100 nm, for organics it is around 50 nm). This suggests that amorphous inorganic absorbers can be used while maintaining a reasonable filling factor. Alternatively, solution processed polymers present an ideal route to easy fabrication and most of our preliminary work has been carried out using poly 3 hexythiophene (P3HT:PCBM), including the first optical confinement geometry solar cell [53].

Most organic materials have larger band gaps ($>$1.9 eV), and only a part of the incident solar light is absorbed. Recently, researchers have reported stable, small band gap conjugated polymers [54, 55], such as poly (5,7-bis(4-decanyl-2-thienyl)-thieno(3,4-b)diathiazole-thiophene-2,5) (PDDTT) with a spectral response from 300 to 1,450 nm. These materials and others [56–58], are promising, but reported efficiencies are still under 10%. Certainly it is reasonable that these performances will rise as processing and bulk heterojunction formation dynamics become better understood for the system. However, there remain some issues in the use of such materials which will remain difficult to answer. Specifically, by using the donor-acceptor structure to reduce band gap and broaden spectral overlap with solar, the density of states across this energy window is lowered. This is essentially because the HOMO bands formed by the donor and the LUMO bands formed by the acceptor do not "add" electron states significantly over their wide band gap analogues. Thus, these materials may present a lower optical density and thicker films must be used to effectively absorb the light. Generally speaking, this will require that the overall mobility of the carriers be increased proportionally and in a balanced fashion, because thicker films will lead to space charge buildup and/or high radiative recombination losses [59]. At the same time, for most standard laboratory devices (planar spun casting devices), there are lots of other places where energy is lost: in the, ITO, PEDOT, and even the Al.

The oscillator strength (optical density) problem leads to a second issue faced by most planar devices. The mobility of the excitons, and the mobility of the carriers usually require the absorbing layer to be made very thin. This reduces exciton recombination and allows removal of separated excitons such that the space charge near the contacts is minimized. However, since the back place of the photoactive absorber is a metal contact (a mirror), the electric field in the absorber (near the contact) is going to zero. Consequently, much of the radiation is reflected away. As we will discuss below, this is known as the "thin film" effect, and is well known and calculated by several groups [21, 60, 61].

In addition, the low mobility of the charge carriers themselves presents problems in achieving performance. Even when the charge mobility is well balanced, the relatively low mobility of the carriers leads inevitably to space charge near both electrodes. This comes from the finite time taken to transfer charge from the organic conductor to the metal/ITO contacts as well as the relatively slow replenishment of charge as current is extracted. This added "internal resistance" must always limit the total power generated by the device relative to its optimum power (also known as the filling factor).

Therefore, for getting higher efficiencies in organic photovoltaics, it may not be enough to synthesize new materials. Strategies may also involve ways of mitigating the effects of optical density as well as of space charge and low mobility. Specifically, several approaches to engineering organic photovoltaics for better optical performance have been explored.

1.3.1 Several Approaches to Improve Optical Performance of Organic Photovoltaics

Recently, an optical spacer [16] has been shown to raise efficiency in organic solar cells by allowing the layered structure to avoid destructive interference within the charge separating layer [56]. As mentioned above, the optical interference between the incident (from the ITO side) and back-reflected light, the intensity of the light in the charge separation layer is reduced near the metallic (Al) electrode. This "optical interference effect" is especially important for thin-film structures where layer thicknesses are comparable to the absorption depth and the wavelength of the incident light, as is the case for photovoltaic cells fabricated from semiconducting polymers. In addition, the use of TiOx as the spacer material can also act as a good hole blocking layer [62], since the top of the valence band of TiOx is at nearly 8.1 eV. Figure 1.17 compares the optical field strength distribution between devices with the TiOx optical spacer and a conventional cell.

For example, Dr. Zhao [63] has used MoO_3 as optical spacer to focus the optical field on the absorber layer P3HT:PCBM and improve the absorption. In Fig. 1.18, it can be seen that the intensity maximum resides close to the center of the active layer. However, the relative optical intensity decreases with the increase of MoO_3 thickness. This means that the thickness of MoO_3 changes the distribution of optical field and corresponding energy. Thus, it also provides a way to reduce the interference and reflection effects associated with the thin film effect.

The tandem solar cell is another way to improve the light absorption and Voc, using several stacked sub-cells with different band gaps. The semiconductors with different bandgaps enable absorption over a broad range of photon energies within the solar emission spectrum; The cells typically utilize a wide bandgap semiconductor for the top cell (first incidences) and a smaller bandgap semiconductor for the bottom cell (when light finally enters) [64]. Using this stacked architecture,

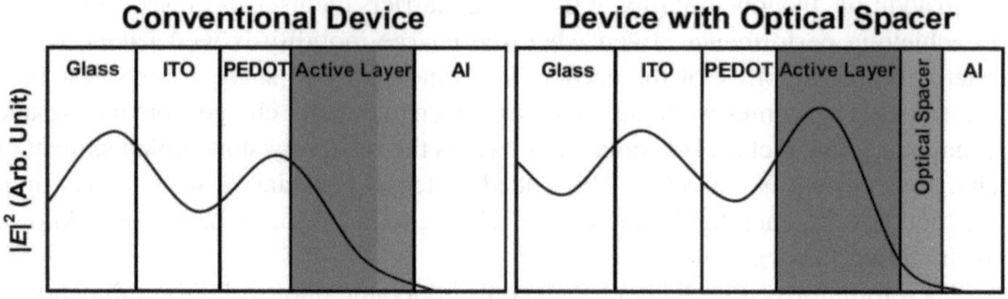

Fig. 1.17 Schematic representation of the spatial distribution of the squared optical electric field strength $|E|^2$ inside the devices with a structure of ITO/PEDOT/active layer/Al (*left*) and ITO/PEDOT/active layer/optical spacer/Al (*right*) (Reprinted with permission from Jin Young Kim, Sun Hee Kim, Hyun-Ho Lee, Kwanghee Lee, Wanli Ma, Xiong Gong, Alan J. Heeger. New architecture for high-efficiency polymer photovoltaic cells using solution-based titanium oxide as an optical spacer. Adv. Mater. **18**, 572–576 (2006); Copyright John Wiley and Sons)

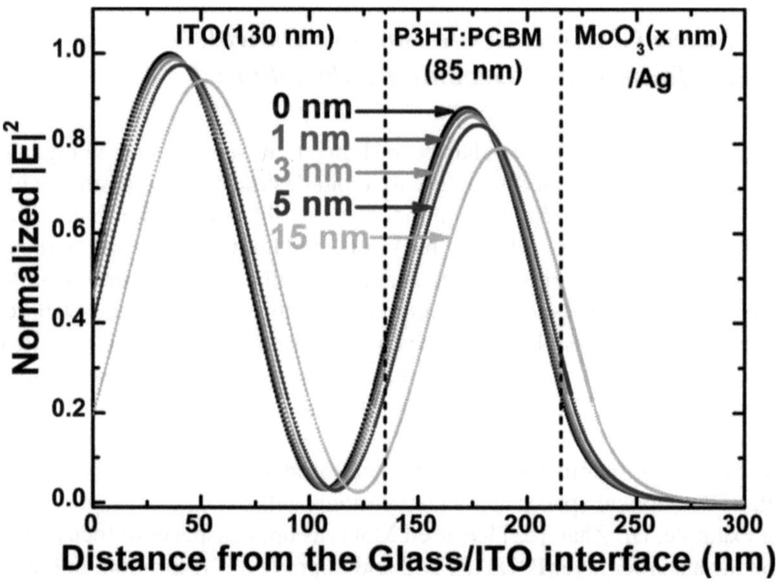

Fig. 1.18 The simulated optical field distribution (for 520 nm illumination) as a function of the distance from ITO/P3HT:PCBM interface in these inverted cells. These cells have the structures of ITO/Ca(1 nm)/P3HT:PCBM(85 nm)/MoO_3(x nm)/Ag(100 nm) with x = 0, 1, 3, 5, and 15. The ultrathin Ca is neglected in the simulation (Reprinted with permission from D. W. Zhao, P. Liu, X. W. Sun, S. T. Tan, L. Ke, A. K. K. Kyaw. An inverted organic solar cell with an ultrathin Ca electron-transporting layer and MoO_3 hole-transporting layer. Appl. Phys. Lett. **95**, 153304 (2009). Copyright 2009, American Institute of Physics)

it is possible to make a thinner absorber layers to avoid high recombination rate and keeping high absorption by several sub-cells with different band gaps. Two methods are available to stack these cells: parallel or serial connections.

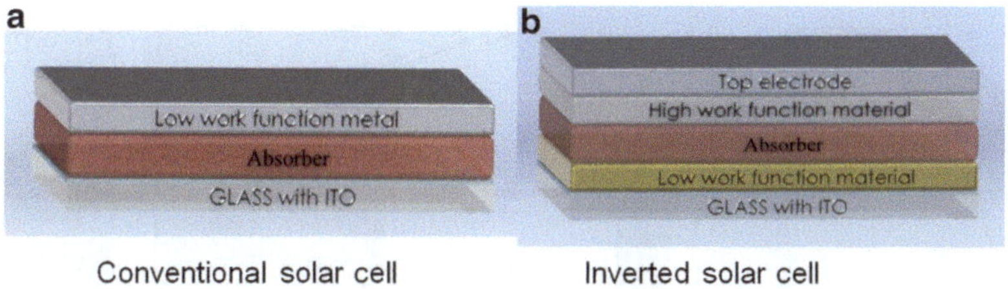

Fig. 1.19 Architecture comparison of conventional solar cell and inverted solar cell

Serial connection means adding up the open circuit voltage of the cells, and parallel connection adds up the short circuit current of the cells. Roughly calculating, the short circuit in serial connection is determined by the lower current of the cells, and the Voc in parallel connection is driven by the lower cell voltage. However, as discussed in the Sect. 1.2.4.1.2, this assumption neglects the influences from series resistance, parallel resistance, and the saturated current from the diode in equivalent circuit, as shown in Fig. 1.12.

In addition, inverted structural organic solar cells (OSCs), which have attracted much attention in recent years, use a material (metal or work metal oxide) with high work function as the top electrode, and use a material with low work function as the bottom electrode. The inverted configuration OSCs with high work function metal as top cathode could improve OSCs' lifetime, i.e., protecting cells from damage by oxygen and moisture in air. Furthermore, the inverted configuration OSCs are the appealing alternative to the conventional regular structure due to the inherent vertical phase separation in the polymer active layers and high stability or long device lifetime [65]. By further integrating it with tandem cells, it is possible to make high performance OSCs with 15% efficiency Fig. 1.19 [66].

However, optical spacers, tandem architectures, and inverted configurations or the combination with tandem designs all include a special layer, such as TiO_x [67], MoO_3 [68], or ZnO [69]. These layers would lead to extra light absorption and energy loss. Therefore, a question comes: can we design a solution to reach high efficiency without inserting extra layers?

1.3.2 Architectures of Optical Confinement Geometry Solar Cell

Another way of stating the above limitations on all photovoltaic devices, including organics, in reaching their theoretical maximum, is that devices are not capable of both efficient charge transport and optimal optical absorption for a given absorption band. For example, for the P3HT:PCBM devices, the ideal thickness for absorption is greater than 250 nm. However, transport of photogenerated excitons and charged polarons in the device is dominated by hopping mechanisms, requiring the active layer thickness to be less than 100 nm for balanced and efficient charge

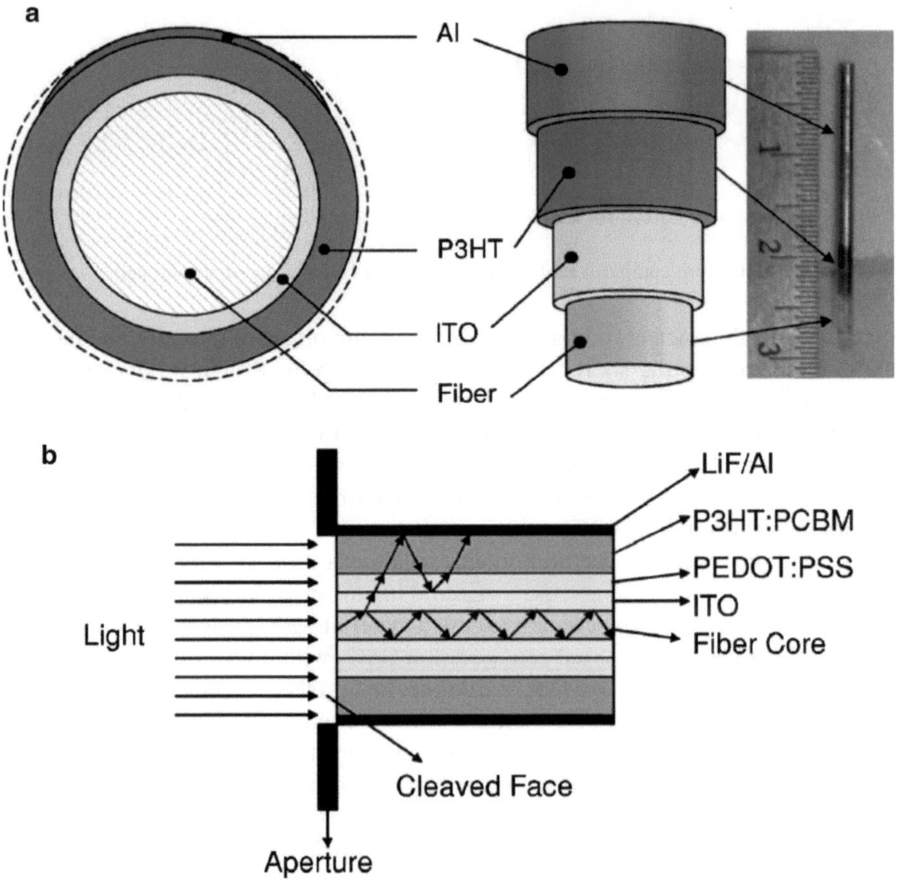

Fig. 1.20 Schematic of the early fiber photovoltaic cell architecture (Reprinted with permission from Jiwen Liu, Manoj A. G. Namboothiry, and David L. Carroll, Appl. Phys. Lett, 90, 133515 (2007) Copyright 2007, American Institute of Physics)

removal even when bulk heterojunctions are employed. Unfortunately, the device cannot be both [44].

Ideally, the radiant energy could be "waveguided" into organic thin film devices such that reflective and transmissive losses were minimized while keeping the electronic properties the same, such that the efficiencies could be raised significantly. We have recently reported on an approach to doing exactly this, using fiber-based solar cell geometries. We have examined the geometric optical considerations of coupling light into thin, organic active layers from a waveguiding fiber (Fig. 1.20). Utilizing standard multimode optical fibers, we have fabricated thin film devices which we refer to as "claddings" based on the bulk heterojunction blend P3HT: PCBM as an absorbing material and indium tin oxide ITO as the transparent conductor. Surprisingly, this architecture works with active film thicknesses significantly greater than those typical for thin film devices, suggesting

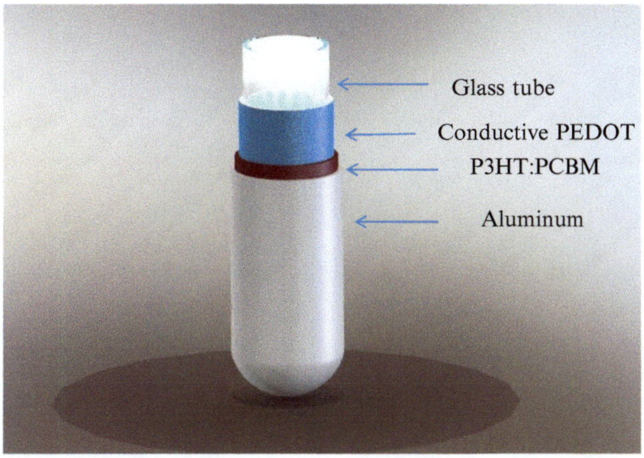

Fig. 1.21 Configuration of a tube based dye-sensitized solar cell, the light illumination at the open end of the tube

that propagating modes within the layer are playing a role in the performance of the device. The experiment shows fiber-based devices are a reasonable approach to creating building blocks for higher performance organic platforms [46]. This has recently been repeated by several groups using dye-sensitized cells as the organic absorbing system [70].

To avoid the reflection on the upper end surface and combine this optical confinement geometry with solution-based solar cells, we demonstrated a waveguiding tube-based optical geometry that provides a general approach to improving organic photovoltaic performance [71]. By fabricating bulk-heterojunction photovoltaics onto thin tubular light pipes, the optical energy can be effectively captured within the absorbing layer without associated reflective losses at the front and rear surfaces of the devices as is typical in planar structures. As shown in Fig. 1.21, we have used a common absorber system: poly-3-hexyl-thiophene:phenyl-C61-butyric-acid-methyl-ester to demonstrate these tubular optical geometries result in very little overall radiative loss. Surprisingly, this also leads to an overall broadening of the absorption window for the device as indicated by the external quantum efficiency (details see Sect. 5.2).

We then invented the oil filled tubular optical concentrators coupled with a model organic bulk hetrerojunction photovoltaic: poly-3-hexathiophene-[6]-phenyl-C61-butyric-acid methyl-ester(P3HT:PCBM) to create a photovoltaic-photothermal hybrid solar collector,[72] as shown in Fig. 1.21. The organic photovoltaic cells were fabricated onto on half of a tubular light pipe and then silicone oil was inserted inside the pipe. This allows solar energy in the visible wavelengths to be effectively converted into electricity by a photocell, while simultaneously, the silicone oil captures the infrared radiation (IR) part of the spectrum as heat energy. The VIS-IR power conversion efficiency for this model organic system, under normal incident AM1.5 G illumination was found to be: PCE ~ 28%, which is combined by

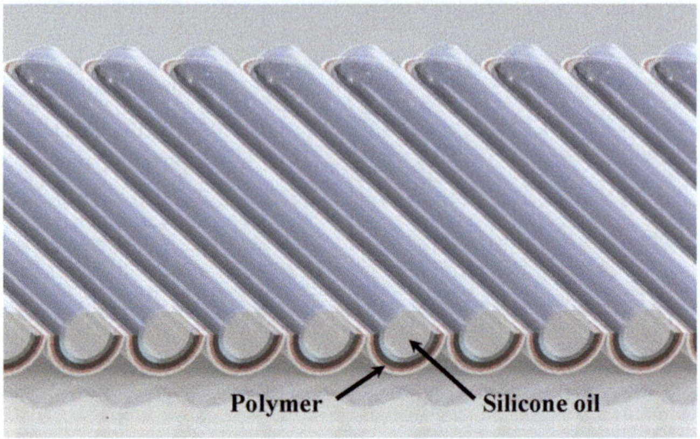

Fig. 1.22 Large area photovoltaic-thermal solar energy collectors based on optical tubes. *Red* is the polymer absorbing light; *White* area in the center of tube is the silicon oil

the photovoltaic efficiency (PCE ~ 2%) and the photothermal efficiency (PCE ~ 26%). We further show that the oil-filled tube acts as a passive optical element that concentrates the light onto the photovoltaic and thereby increases its overall efficiency Fig. 1.22 (details in Sect. 7.1).

If silicone oil is replaced by dye solution with a different band gap, it will convert light to current directly instead of through thermal intermediation. Therefore, the tube based architecture was investigated as a light-trap and optical guider to enhance light absorption in DSSC photovoltaic devices in Fig. 1.23. We also simulated the effect of incident light angle to the absorption of a tube based DSSC (TBDSSC). The incident angle exhibited a definitive effect on the light harvesting of a TBDSSC and thus determined the photocurrent of a TBDSSC. We also studied the influence of photoanode length and thickness to the performances of the TBDSSC. We get a maximum efficiency of 2.54% with a 3 mm porous TiO_2 layer (PTL) length, 3.3 μm PTL thickness and 5° incident light angle. The fabricated single device presents a high surface-volume ratio, which indicated that the tube based DSSC is a promising candidate for high efficiency PV devices (details in Sect. 7.2).

To make a large area device, we first reported the fiber-based solar cell on stamped polycarbonate substrates prepared by Diamond Films in Auckland NZ using a proprietary stamping process. Apart from this method, the laser processing was also used to fabricate the fiber array substrate. Figure 1.24 shows the devices fabricated by spray coating method, with architecture ITO/PEDOT/P3HT:PCBM/Al, for two types of devices (details in Sect. 6.2).

In sum, we classify the optical confinement geometry solar cell as a single device, integrated device and hybrid device in terms of their structures and features, which will be discussed in detail from Chaps. 5 to 7.

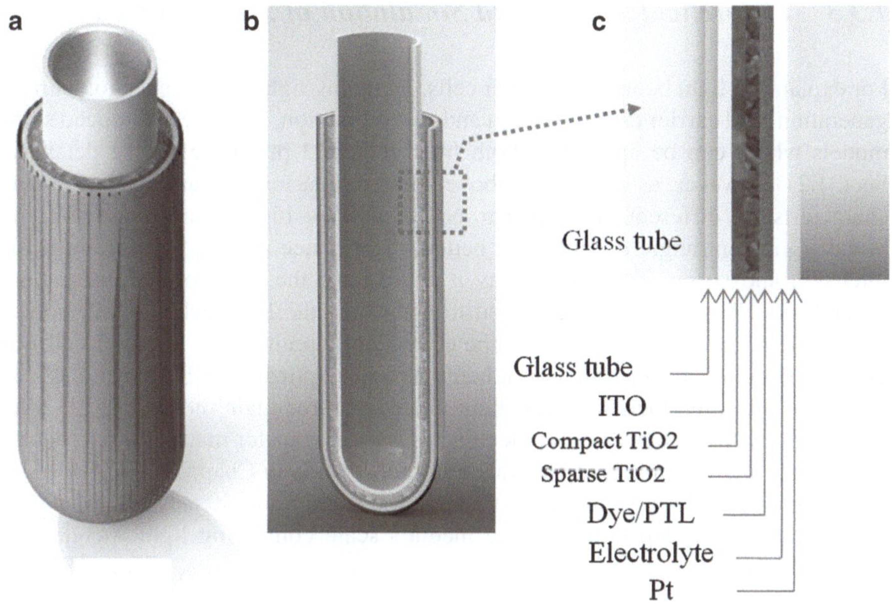

Fig. 1.23 (**a**) Architecture of a tube based dye-sensitized solar cell. (**b, c**) Is the corresponding cross section graph and partial enlargement of the TBDSSC structure

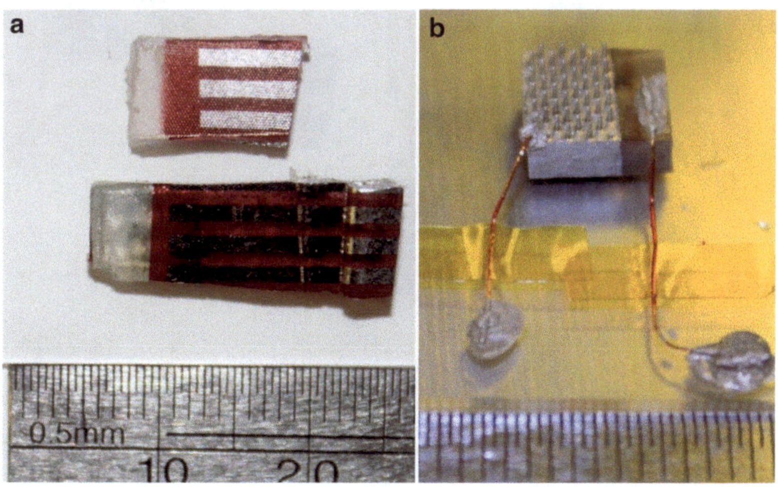

Fig. 1.24 Fiber-solar cells on two types of stamped substrates, (**a**) is a stamped dome device, (**b**) is a fiber array substrate

1.3.3 Theoretical Models and Simulation of 3D Solar Cells

For explaining light behaviors in fiber cells, including light incidence, photocurrent generation, and carrier drift, diffusion and recombination, we have introduced some models which can be applied to both fiber cells and planar cells (see details in Sect. 1.2). However, as mentioned above, three dimensional solar cells have many characteristics different from planar cells, such as high absorption spectrum, broaden incident angle dependence, better performance at very high illumination intensity, and so on. To explain why it could have these performances, another several models need to be introduced to determine the light behaviors in optical confinement geometry. In addition, the equivalent circuit of the 3D solar cell was given to simulate its performance at heterogeneous optical distribution and at very high light intensity. Generally speaking, there are three main models for 3D solar cells, namely ray tracing model, (details in Chap. 3) transfer matrix model (details in Chap. 4) and the model of equivalent circuit (details in Chap. 2).

As shown in Fig. 1.25, light is studied by assuming it as an optical line or electrical wave depending on the media's scale comparing light wavelength.

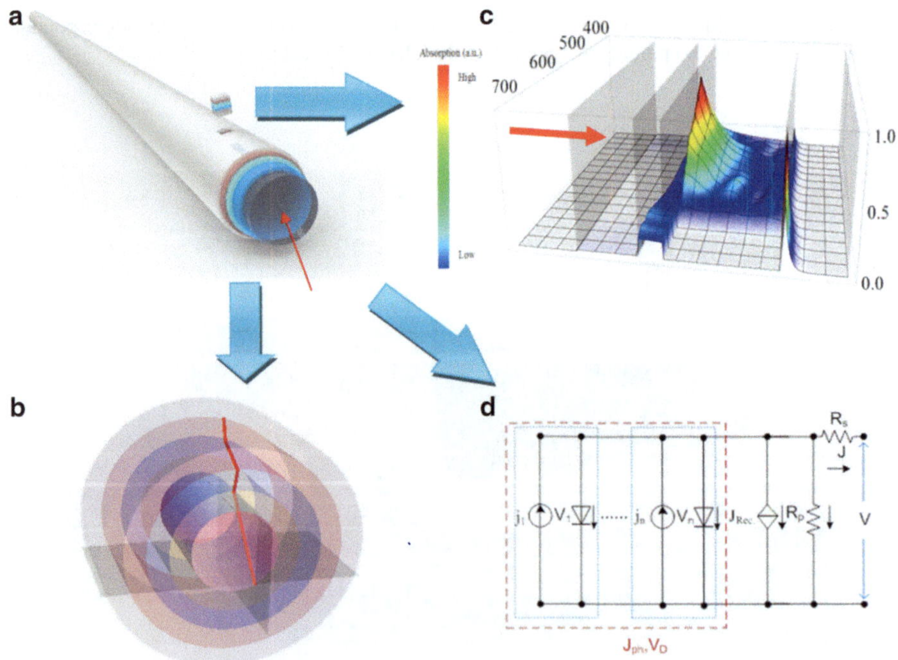

Fig. 1.25 (**a**) Shows a combined scheme of three main models in optical confinement geometry. Separately, the light behaviors in fiber cavity is described by ray tracing model (**b**), that in multi-layers is explained by transfer matrix model (**c**), and that whole electrical performance is shown in the equivalent circuit in (**d**)

While light enters into a fiber cavity with micrometer or millimeter diameter and centimeter length in Fig. 1.25c, light can be described as some optical lines. By means of the ray tracing model for optical lines, it can predict light absorption influenced by the coupling angle into the fiber, the fiber diameter and length. When it propagates in multi-layers in the scale of nanometers (~100 μm, relative to the wavelengths of visible light ~100 nm), light mainly exhibits the nature of electrical wave and has interference between layers. This could be evidence that the mode structure associated with a fiber device leads to an overall shift and broadening of the external quantum efficiency (EQE) for any given polymer absorber, and thickness dependence of current density by further combing the model of carrier drift, diffusion, and recombination. The third model provides a method to predict the output by heterogeneous optical distribution or at high incident light intensity.

These three models of optical confinement geometry solar cells will be discussed in detail in Chaps. 2, 3and 4. Afterwards, based on these models, three series of optical confinement geometry solar cells are introduced in Chaps. 5, 6, and 7, about their special characteristics and improved performances.

References

1. P. Chen, *Civilization Bifurcation, Economic Chaos, and Evolutionary Economic Dynamics* (Peking University Press, Beijing, 2004), p. 55
2. German Advisory Council on Global Change (WBGU), *Towards Sustainable Energy Systems* (Earthscan, London, 2003), p. 242
3. S. Guha, J. Yang, A. Banerjee, Amorphous silicon alloy photovoltaic research – present and future. Prog. Photovoltaics **8**, 141–150 (2000)
4. R. Noufi, H.W. Schock, CIGS-based solar cells for the next millennium. Prog. Photovoltaics **8**, 151–160 (2000)
5. D. Bonnet, P. Meyers, Cadmium-telluride – material for thin film solar cells. J. Mater. Res. **13**, 2740–2753 (1998)
6. R. Singh, P.C. Chou, U. Purkayastha, Recent development in gaas and other Iii-V-semiconductor solar-cells grown by metallorganic chemical vapor-deposition (Mocvd). J. Electrochem. Soc. **132**, C217–C218 (1985)
7. M. Gratzel, Perspectives for dye-sensitized nanocrystalline solar cells. Prog. Photovoltaics **8**, 171–185 (2000)
8. A.P. Alivisatos, W.U. Huynh, J.J. Dittmer, Hybrid nanorod-polymer solar cells. Science **295**, 2425–2427 (2002)
9. http://en.wikipedia.org/wiki/Sunlight
10. ASTM Standard G173, 2003e1: Standard tables for reference solar spectral irradiances: Direct normal and hemispherical on 37° tilted surface (2003). ASTM International, West Conshohocken, PA doi: 10.1520/G0173-03E01, www.astm.org
11. M.A. Green, *Third Generation Photovoltaics: Advanced Solar Energy Conversion* (Springer, Berlin/New York, 2003)
12. T. Trupke, M.A. Green, P. Wurfel, Improving solar cell efficiencies by down-conversion of high-energy photons. J. Appl. Phys. **92**, 1668–1674 (2002)
13. R.R. King, D.C. Law, K.M. Edmondson, C.M. Fetzer, G.S. Kinsey, H. Yoon, R.A. Sherif, N.H. Karam, 40% efficient metamorphic GaInP/GaInAs/Ge multijunction solar cells. Appl. Phys. Lett. **90**(18), 183516 (2007), AIP

14. S.R. Cowan, A. Roy, A.J. Heeger, Recombination in polymer-fullerene bulk heterojunction solar cells. Phys. Rev. B. **82**, 245207 (2010)
15. J.K. Lee, N.E. Coates, S. Cho, N.S. Cho, D. Moses, G.C. Bazan, K. Lee, A.J. Heeger, Efficacy of TiO[sub x] optical spacer in bulk-heterojunction solar cells processed with 1,8-octanedithiol. Appl. Phys. Lett. **92**, 243308 (2008)
16. J.Y. Kim, S.H. Kim, H.-H. Lee, K. Lee, W. Ma, X. Gong, A.J. Heeger, New architecture for high-efficiency polymer photovoltaic cells using solution-based titanium oxide as an optical spacer. Adv. Mater. **18**, 572–576 (2006)
17. D.W. Zhao, L. Ke, Y. Li, S.T. Tan, A.K.K. Kyaw, H.V. Demir, X.W. Sun, D.L. Carroll, G.Q. Lo, D.L. Kwong, Optimization of inverted tandem organic solar cells. Sol. Energ. Mat. Sol. C., (Corrected Proof, 2010 in press)
18. A.K. Ghosh, T. Feng, Merocyanine organic solar-cells. J. Appl. Phys. **49**, 5982–5989 (1978)
19. C.W. Tang, A.C. Albrecht, Photovoltaic effects of metal-chlorophyll-a-metal sandwich cells. J. Chem. Phys. **62**, 2139–2149 (1975)
20. A.K. Ghosh, D.L. Morel, T. Feng, R.F. Shaw, C.A. Rowe, Photovoltaic and rectification properties of Al-Mg phthalocyanine-Ag Schottky-barrier cells. J. Appl. Phys. **45**, 230–236 (1974)
21. L.A.A. Pettersson, L.S. Roman, O. Inganas, Modeling photocurrent action spectra of photovoltaic devices based on organic thin films. J. Appl. Phys. **86**, 487–496 (1999)
22. D.W. Sievers, V. Shrotriya, Y. Yang, Modeling optical effects and thickness dependent current in polymer bulk-heterojunction solar cells. J. Appl. Phys. **100**, 114509 (2006)
23. F. Richter, M. Florian, K. Henneberger, Poynting's theorem and energy conservation in the propagation of light in bounded media. Europhys. Lett. **81**, 67005 (2008)
24. L.J.A. Koster, E.C.P. Smits, V.D. Mihailetchi, P.W.M. Blom, Device model for the operation of polymer/fullerene bulk heterojunction solar cells. Phys. Rev. B. **72**, 085205 (2005)
25. S. Selberherr, *Analysis and Simulation of Semiconductor Devices* (Springer, Wien, 1984)
26. C.L. Braun, Electric-field assisted dissociation of charge-transfer states as a mechanism of photocarrier production. J. Chem. Phys. **80**, 4157–4161 (1984)
27. H.K. Gummel, A self-consistent iterative scheme for one-dimensional steady state transistor calculations. IEEE T. Electron. Dev. **11**, 455–465 (1964)
28. P. Würfel, U. Würfel, in Physics of Solar Cells : From Basic Principles to Advanced Concepts, 2nd, updated and expanded edn. (Wiley-VCH, Weinheim, 2009)
29. A. Jain, A. Kapoor, A new approach to study organic solar cell using Lambert W-function. Sol. Energ. Mat. Sol. C. **86**, 197–205 (2005)
30. R.H. Bube, A.L. Fahrenbruch, *Advances in Electronics and Electron Physics* (Academic, New York, 1981), p. 163
31. A.L. Fahrenbruch, J. Aranovich, *Solar Energy Conversion* (Springer, New York, 1979), p. 257
32. B.P. Rand, D.P. Burk, S.R. Forrest, Offset energies at organic semiconductor heterojunctions and their influence on the open-circuit voltage of thin-film solar cells. Phys. Rev. B. **75**, 115327 (2007)
33. Y. Li, M. Wang, H. Huang, W. Nie, Q. Li, E.D. Peterson, R. Coffin, G. Fang, D.L. Carroll, Influence on open-circuit voltage by optical heterogeneity in three-dimensional organic photovoltaics. Phys. Rev. B. **84**, 085206 (2011)
34. G. Li, V. Shrotriya, J.S. Huang, Y. Yao, T. Moriarty, K. Emery, Y. Yang, High-efficiency solution processable polymer photovoltaic cells by self-organization of polymer blends. Nat. Mater. **4**, 864–868 (2005)
35. F.A. Shirland, The history, design, fabrication and performance of CdS thin film solar cells. Adv. Energ. Convers. **6**, 201–221 (1966)
36. K.I. Ishibashi, Y. Kimura, M. Niwano, An extensively valid and stable method for derivation of all parameters of a solar cell from a single current–voltage characteristic. J. Appl. Phys. **103**, 094507 (2008)
37. A. Jain, A. Kapoor, A new method to determine the diode ideality factor of real solar cell using Lambert W-function. Sol. Energ. Mat. Sol. C. **85**, 391–396 (2005)

38. G. Williams, Advances in organic photovoltaics and methods for effective solar cell parameter extraction, http://www.eng.uwaterloo.ca/~g3willia/documents/ECE730_OSCReview.pdf
39. K. Bouzidi, M. Chegaar, A. Bouhemadou, Solar cells parameters evaluation considering the series and shunt resistance. Sol. Energ. Mat. Sol. C. **91**, 1647–1651 (2007)
40. A. Ortiz-Conde, F.J.G. Sanchez, J. Muci, New method to extract the model parameters of solar cells from the explicit analytic solutions of their illuminated I-V characteristics. Sol. Energ. Mat. Sol. C. **90**, 352–361 (2006)
41. N. Nehaoua, Y. Chergui, D.E. Mekki, Determination of organic solar cell parameters based on single or multiple pin structures. Vacuum **84**, 326–329 (2009)
42. Z. Ouennoughi, M. Chegaar, A simpler method for extracting solar cell parameters using the conductance method. Solid State Electron. **43**, 1985–1988 (1999)
43. W. Shockley, H.J. Queisser, Detailed balance limit of efficiency of p-n junction solar cells. J. Appl. Phys. **32**, 510-519 (1961), AIP
44. J.W. Liu, M.A.G. Namboothiry, D.L. Carroll, Fiber-based architectures for organic photovoltaics. Appl. Phys. Lett. **90**, 063501 (2007)
45. Y. Li, W. Zhou, D. Xue, J.W. Liu, E.D. Peterson, W.Y. Nie, D.L. Carroll, Origins of performance in fiber-based organic photovoltaics. Appl. Phys. Lett. **95**, 203503 (2009)
46. J.W. Liu, M.A.G. Namboothiry, D.L. Carroll, Optical geometries for fiber-based organic photovoltaics. Appl. Phys. Lett. **90**, 133515 (2007)
47. O.L. Muskens, J.G. Rivas, R.E. Algra, E.P.A.M. Bakkers, A. Lagendijk, Design of light scattering in nanowire materials for photovoltaic applications. Nano Lett. **8**, 2638–2642 (2008)
48. J.H. Woodruff, J.B. Ratchford, I.A. Goldthorpe, P.C. McIntyre, C.E.D. Chidsey, Vertically oriented germanium nanowires grown from gold colloids on silicon substrates and subsequent gold removal. Nano Lett. **7**, 1637–1642 (2007)
49. H. Zhou, A. Colli, T. Butler, N. Rupesinghe, A. Mumtaz, G. Amaratunga, J.I.B. Wilson, Carbon nanotube arrays for optical design of amorphous silicon solar cells. Int. J. Mater. Form. **1**, 113–116 (2008)
50. A. Barnett, D. Kirkpatrick, C. Honsberg, D. Moore, M. Wanlass, K. Emery, R. Schwartz, D. Carlson, S. Bowden, D. Aiken, A. Gray, S. Kurtz, L. Kazmerski, M. Steiner, J. Gray, T. Davenport, R. Buelow, L. Takacs, N. Shatz, J. Bortz, O. Jani, K. Goossen, F. Kiamilev, A. Doolittle, I. Ferguson, B. Unger, G. Schmidt, E. Christensen, D. Salzman, Very high efficiency solar cell modules. Prog. Photovoltaics **17**, 75–83 (2009)
51. S. Curran, J. Talla, S. Dias, J. Dewald, Microconcentrator photovoltaic cell (the m-C cell): Modeling the optimum method of capturing light in an organic fiber based photovoltaic cell. J. Appl. Phys. **104**, 064305 (2008)
52. B. Myers, M. Bernardi, J.C. Grossman, Three-dimensional photovoltaics. Appl. Phys. Lett. **96**, 071902 (2010)
53. Z.X. Liu, A. Masuda, T. Nagai, T. Miyazaki, M. Takano, M. Takano, H. Yoshigahara, K. Sakai, K. Asai, M. Kondo, A concentrator module of spherical Si solar cell. Sol. Energ. Mat. Sol. C. **91**, 1805–1810 (2007)
54. X. Gong, M. Tong, Y. Xia, W. Cai, J.S. Moon, Y. Cao, G. Yu, C.-L. Shieh, B. Nilsson, A.J. Heeger, High-detectivity polymer photodetectors with spectral response from 300 nm to 1450 nm. Science **325**, 1665–1667 (2009)
55. R.C. Coffin, J. Peet, J. Rogers, G.C. Bazan, Streamlined microwave-assisted preparation of narrow-bandgap conjugated polymers for high-performance bulk heterojunction solar cells. Nat. Chem. **1**, 657–661 (2009)
56. S.H. Park, A. Roy, S. Beaupre, S. Cho, N. Coates, J.S. Moon, D. Moses, M. Leclerc, K. Lee, A.J. Heeger, Bulk heterojunction solar cells with internal quantum efficiency approaching 100%. Nat. Photonics. **3**, 297–U295 (2009)
57. H.Y. Chen, J.H. Hou, S.Q. Zhang, Y.Y. Liang, G.W. Yang, Y. Yang, L.P. Yu, Y. Wu, G. Li, Polymer solar cells with enhanced open-circuit voltage and efficiency. Nat. Photonics. **3**, 649–653 (2009)

58. L.J. Huo, J.H. Hou, H.Y. Chen, S.Q. Zhang, Y. Jiang, T.L. Chen, Y. Yang, Bandgap and molecular level control of the low-bandgap polymers based on 3,6-Dithiophen-2-yl-2,5-dihydropyrrolo [3,4-c]pyrrole-1,4-dione toward highly efficient polymer solar cells. Macromolecules **42**, 6564–6571 (2009)
59. F. Zhang, A. Vollmer, J. Zhang, Z. Xu, J.P. Rabe, N. Koch, Energy level alignment and morphology of interfaces between molecular and polymeric organic semiconductors. Org. Electron. **8**, 606–614 (2007)
60. R. Hausermann, E. Knapp, M. Moos, N.A. Reinke, T. Flatz, B. Ruhstaller, Coupled optoelectronic simulation of organic bulk-heterojunction solar cells: Parameter extraction and sensitivity analysis. J. Appl. Phys. **106**, 104507 (2009)
61. M.G. Harrison, J. Gruner, G.C.W. Spencer, Analysis of the photocurrent action spectra of MEH-PPV polymer photodiodes. Phys. Rev. B. **55**, 7831–7849 (1997)
62. S. Cho, J. Yuen, J.Y. Kim, K. Lee, A.J. Heeger, S. Lee, Multilayer bipolar field-effect transistors. Appl. Phys. Lett. **92**, 063511 (2008)
63. D.W. Zhao, P. Liu, X.W. Sun, S.T. Tan, L. Ke, A.K.K. Kyaw, An inverted organic solar cell with an ultrathin Ca electron-transporting layer and MoO3 hole-transporting layer. Appl. Phys. Lett. **95**, 153304 (2009)
64. M.W. Wanlass, K.A. Emery, T.A. Gessert, G.S. Horner, C.R. Osterwald, T.J. Coutts, Practical considerations in tandem cell modeling. Sol. Cells. **27**, 191–204 (1989)
65. F.J. Zhang, X.W. Xu, W.H. Tang, J. Zhang, Z.L. Zhuo, J. Wang, J. Wang, Z. Xu, Y.S. Wang, Recent development of the inverted configuration organic solar cells. Sol. Energ. Mat. Sol. C. **95**, 1785–1799 (2011)
66. G. Dennler, M.C. Scharber, T. Ameri, P. Denk, K. Forberich, C. Waldauf, C.J. Brabec, Design rules for donors in bulk-heterojunction tandem solar cells-towards 15% energy-conversion efficiency. Adv. Mater. **20**, 579 (2008)
67. T. Kuwabara, H. Sugiyama, M. Kuzuba, T. Yamaguchi, K. Takahashi, Inverted bulk-heterojunction organic solar cell using chemical bath deposited titanium oxide as electron collection layer. Org. Electron. **11**, 1136–1140 (2010)
68. A.K.K. Kyaw, X.W. Sun, C.Y. Jiang, G.Q. Lo, D.W. Zhao, D.L. Kwong, An inverted organic solar cell employing a sol–gel derived ZnO electron selective layer and thermal evaporated MoO(3) hole selective layer. Appl. Phys. Lett. **93**, 221107 (2008)
69. W. Mingjun, L. Yuan, H. Huihui, D.P. Eric, N. Wanyi, Z. Wei, Z. Wei, H. Wenxiao, F. Guojia, S. Nanhai, Z. Xingzhong, L.C. David, Thickness dependence of the MoO3 blocking layers on ZnO nanorod-inverted organic photovoltaic devices. Appl. Phys. Lett. **98**, 103305 (2011)
70. B. Weintraub, Y.G. Wei, Z.L. Wang, Optical fiber/nanowire hybrid structures for efficient three-dimensional dye-sensitized solar cells. Angew. Chem. Int. Edit. **48**, 8981–8985 (2009)
71. Y. Li, E.D. Peterson, H. Huang, M. Wang, D. Xue, W. Nie, W. Zhou, D.L. Carroll, Tube-based geometries for organic photovoltaics. Appl. Phys. Lett. **96**, 243505 (2010)
72. H. Huang, Y. Li, M. Wang, W. Nie, W. Zhou, E.D. Peterson, J. Liu, G. Fang, D.L. Carroll, Photovoltaic-thermal solar energy collectors based on optical tubes. Sol. Energ. **85**, 450–454 (2011)

Chapter 2
Equivalent Circuit

2.1 Equivalent Circuit Model

A particularly attractive feature of such geometries is that it is possible to fabricate an efficient cell using ultra-thin organic thin films (lower than 50 nm), and thereby lead to high filling factors due to decreased recombination (Reprinted with permission from [1], Copyright (2011) by The American Physical Society), [2–4] Further, due to the well-defined mode structure of many of these geometries, there is the possibility of effectively utilizing frequency conversion schemes. Because of the long optical path in the fiber, a frequency convertor can play an effective role in doubling high energy photons to improve current. However, there are still several problems that must be overcome. We have reported that the open-circuit voltage (Voc) tends to decrease in Optical Confinement Geometry Organic Photovoltaics (OCGOPV) geometries [5, 6]. In that earlier work [5], we defined two "active" areas of the general three dimensional geometry as Fig. 2.1a shows: the "Current Active Area" (CAA) is the area from which current is collected and the "Illumination Active Area" (IAA) is the area of illumination of the structure [5]. These are a distinctive feature of any OCGOPV. Essentially for the planar cell, IAA equals CAA, but in the OCGOPV the CAA is much greater than the IAA. In other words, the flux entering the aperture (IAA) is spread over a much larger area within the cell (the CAA) leading to a lower optical intensity on CAA like an inverse concentrator. Because light is generally partitioned into modes of the "confining cavity," the optical intensity in an OCGOPV is typically heterogeneously distributed across the CAA (HeOI), differing from the homogeneous optical intensity (HoOI) in planar OPV. The heterogeneous distribution in OCGOPVs can be simulated by the ray tracing model we reported [7] as shown in Fig. 2.1b. If this power heterogeneity becomes too great, a loss in Voc will occur for the device. In this section, we examine the effects of optical heterogeneity on a model OCGOPV using a composite equivalent circuit analysis.

To understand the performance of an OCGOPV, it is necessary to know the connections between their electrical and optical characteristics. First, as shown in

Y. Li, *Three Dimensional Solar Cells Based on Optical Confinement Geometries*, Springer Theses, DOI 10.1007/978-1-4614-5699-5_2,

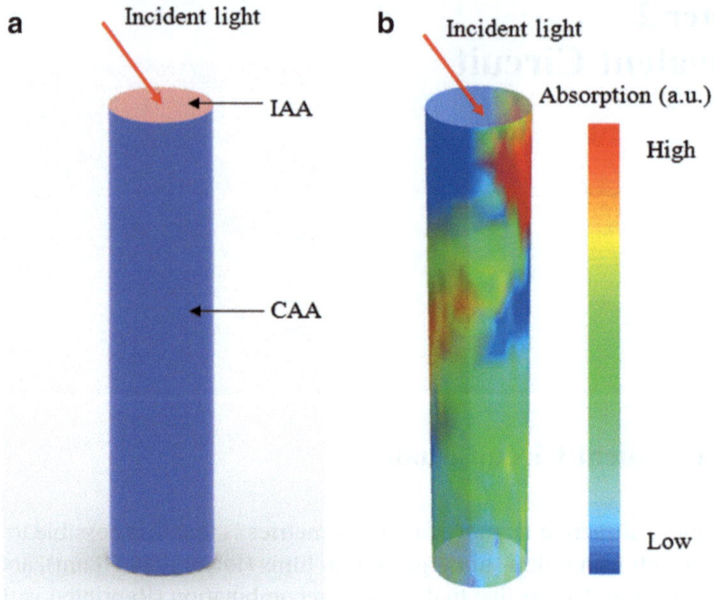

Fig. 2.1 (**a**) IAA and CAA are represented as *pink* area at the top and *blue* area round the fiber. (**b**) Heterogeneous absorption distribution through inner surface in OCGOPVs is simulated by a ray tracing model [7]. The legend at the *right* represents the absorption level at inner surface

Fig. 2.2a, we take a small piece from a whole OCGOPV (a.1) as one *sub*unit (a.2) and treat it like a planar solar cell (a.3). For each *sub*unit in (a.2), when very small, we may assume it fits planar cell theory. In Fig. 2.2b, the equivalent circuit of planar cell is described by J-V characteristics expressed by the generalized Shockley equation [8–10].

$$J = \frac{R_p}{R_s + R_p}\left\{J_s\left[\exp\left(\frac{e(V - JR_s)}{n_D k_B T}\right) - 1\right] + \frac{V}{R_p} - J_{ph} + J_{\text{Rec.}}\right\} \tag{2.1}$$

Where, n_D is the diode ideality factor, J_s is the reverse saturation current, and J_{Rec} is represented by the current source, which counteracts the photocurrent density J_{ph} [11–13]. Due to lower carrier mobility in polymers generally [14–16], the main factors influencing the recombination rate r_{Rec}, are thickness of polymer, the temperature and the fabrication procedure/polymer processing used. For very thin films of bulk heterojunction polymer blends, the influence of $r_{Rec.}$ on J_{sc} can be safely ignored [2].

To account for heterogeneity in the OCGOPVs, the model must consist of many OPV *sub*units with different performances, as shown in Fig. 2.2c. They absorb different photon numbers to contribute many j_i, and each generates a different v_i.

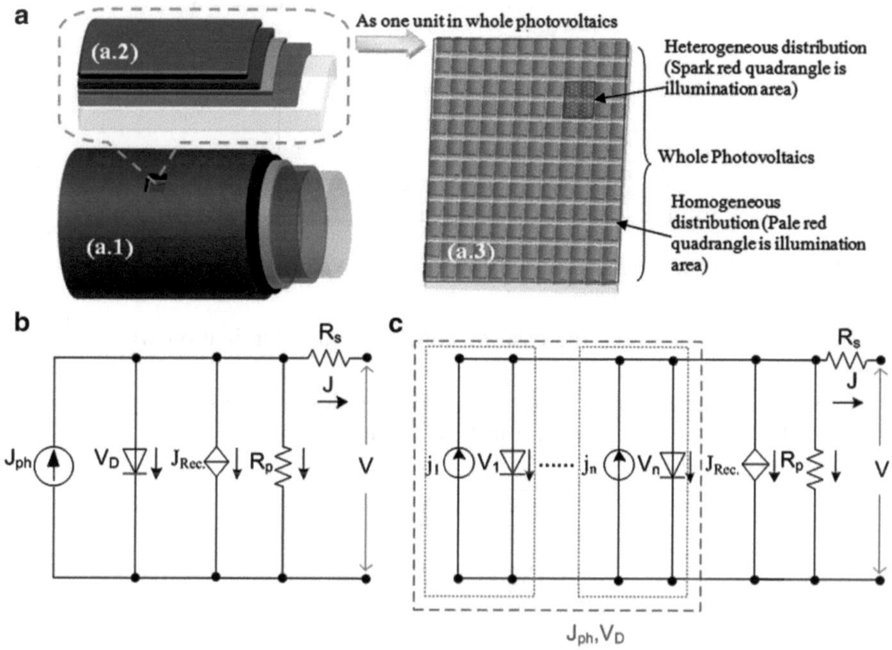

Fig. 2.2 (**a**) From (a.1) to (a.3), our model divides the OCGOPV into *sub*units, and then connects them as a planar photovoltaics. In (a.3), illumination across the *sub*cells is heterogeneously distributed due to the mode structure of the waveguide. (**b**) The equivalent circuit of conventional planar OPV. J_{ph}, $r_{Rec.}$, V_D, R_s, and R_p are photocurrent source, recombination, diode voltage, series resistances and parallel resistances, respectively. J and V are the output current density and voltage of OPV. (**c**) The equivalent circuit of the OCGOPVs composed of unit *sub*cells

The contributions from the individual cells are summed to provide a whole J_{ph} and V_{D} as show below.

$$J_{ph} = \frac{1}{n}\sum_{i=1}^{n} j_i = \frac{1}{n}\sum_{i=1}^{n}\int \frac{q\lambda}{hc}\eta(\lambda)Q_i(\lambda)d\lambda \tag{2.2}$$

Where, $Q_i(\lambda)$ is the spectral irradiance on the active area from the incident light at each *sub*unit, which can be obtained by a 3D fiber-cell model based on ray tracing [7] coupled with a transfer matrix method which accounts for the materials absorption properties [2, 3, 17, 18]. In Eq. 2.2, h, c, q, and λ are the Planck's constant, light speed, electron charge, and wavelength. $\eta(\lambda)$ is the external quantum efficiency of wavelength λ. From Eq. 2.1, when $J = 0$, the open-circuit voltage $V_{oc} = V = V_D$. Also since $J_sR_p >> V_{oc}$ [12], Eq. 2.2 can be simplified as:

$$V_{oc} = \frac{n_D k_B T}{q}\ln\left(\frac{(R_p + R_s)J_{ph}}{J_s R_p} + 1\right), J_{ph} = \frac{1}{n}\sum_{n} j_i \tag{2.3}$$

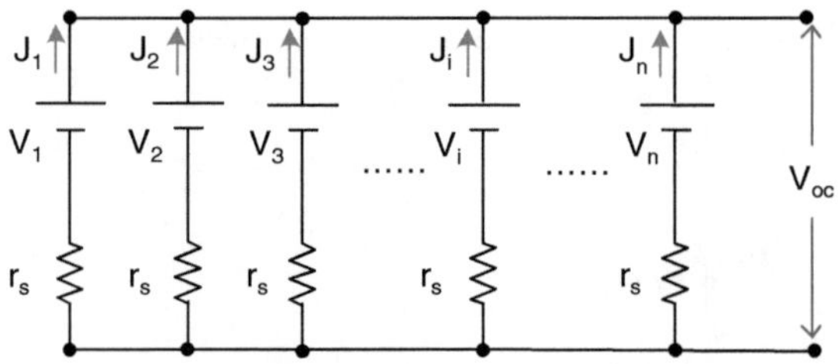

Fig. 2.3 The circuit of superimposed V_i to whole voltage Voc

Equation 2.3 represents the case of homogeneous illumination when all j_i are equal.

In Fig. 2.3, the superimposed voltage V_{oc} is described as the average of all voltage of *sub*cells cells, as shown in Eqs. 2.4 and 2.5.

$$0 = \sum_n j_n = \sum_n \frac{V_{oc} - V_i}{r_s} \tag{2.4}$$

Thus,

$$\mathrm{V_{oc}} = \frac{\sum\limits_n \mathrm{V_i/r_s}}{\sum\limits_n 1/\mathrm{r_s}} = \sum_n \mathrm{V_i} \tag{2.5}$$

$$V_{oc} = \frac{1}{n}\sum_n V_i \tag{2.6}$$

$$V_i = \frac{n_D k_B T}{q} \ln\left(\frac{(R_p + R_s)j_i}{J_s R_p} + 1\right) \tag{2.7}$$

Therefore, when the optical intensity is heterogeneous, i.e. each *sub*unit cell absorbs different light fluxes and contributes various voltages, their superimposed voltage V_{oc} is described as the average of all voltage of unit *sub*cells in Eq. 2.6 [19].

When j_i is constant and equals the average current density of the whole cell, Eq. 2.7 becomes Eq. 2.3 of the homogeneous case. Actually, j_i is a function of position x, namely $j_i = j(x)$, which is related to optical distribution by Eq. 2.2. We can use the method of calculus of variations to express the relationship of V_{oc} with optical distribution $j_i(x)$. We assume

$$V_{oc} = \frac{1}{n}\sum_n a\ln(bj_i(x) + 1) \tag{2.8}$$

$$\text{where } \Psi\mathrm{a} = \frac{\mathrm{n_D k_B T}}{\mathrm{q}},\ \mathrm{b} = \frac{\mathrm{R_p + R_s}}{\mathrm{J_s R_p}} \tag{2.9}$$

Equation 2.8 is rewritten as an integral.

$$V_{oc} = \frac{1}{n}\sum_n a\ln(bj_i(x) + 1) = \int_0^{x_0} \frac{a}{x_0}\ln(bj(x) + 1)\mathrm{d}x \tag{2.10}$$

Where, x_0 is the area of whole cell. And we assume the total current is constant, and the range of x_i is invariant, i.e. the area x_0 is constant. Then, the effect of distribution of $j(x)$ can be studied n terms of the definition of functional derivative

$$\delta F = \int d^3 r d \frac{\delta \mathrm{F}}{\delta \mathrm{n(r)}} \delta \mathrm{n(r)} \tag{2.11}$$

$$\begin{aligned}
\delta V_{oc} &= \int dx d \frac{\delta \mathrm{V_{oc}}}{\delta \mathrm{j(x)}} \delta \mathrm{j(x)} \\
&= \frac{a}{x_0} \int_0^{x_0} \{\ln(b[j(x) + \delta j(x)] + 1) - \ln(bj(x) + 1)\} \mathrm{dx} \\
&= \frac{a}{x_0} \int_0^{x_0} \{\ln(bj(x) + 1) + \frac{b\delta j(x)}{bj(x) + 1} - \ln(bj(x) + 1)\} \mathrm{dx} \\
&= \frac{a}{x_0} \int_0^{x_0} \frac{b\delta j(x)}{bj(x) + 1} \mathrm{dx}
\end{aligned} \tag{2.12}$$

The kernel of that integral determines the influence of the variation of distribution $j(x)$. By put Eq. 2.9 and the following parameters into this kernel.

$$\begin{aligned}
T &= 300K \\
n_D &= 1 \\
k_B &= 1.3806505 \times 10^{-23} J/K \\
q &= 1.6 \times 10^{-19} C
\end{aligned} \tag{2.13}$$

Also, when j_s is much less than $j(x)$, the kernel becomes

$$\begin{aligned}
\frac{a}{x_0} \frac{b}{bj(x) + 1} &= \frac{0.026V}{x_0} \frac{r_p + r_s}{j_s r_p + j(x)(r_p + r_s)} \\
&\approx \frac{0.026V}{x_0} \frac{r_p + r_s}{j(x)(r_p + r_s)} \\
&= \frac{0.026V}{x_0} \frac{1}{j(x)}
\end{aligned} \tag{2.14}$$

Where, x_0 can be assumed as a unit area, $j(x)$ can be regard as the average current density in homogeneous, and $\delta j(x)$ is the perturbation in heterogeneous distribution. This means that the heterogeneous illumination can be described by the homogeneous model if the ratio $\delta j(x)/j(x)$ is of the same order of magnitude. Thus, Eq. 2.12 becomes

$$\delta V_{oc} \approx \int_0^{x_0} \frac{0.026V}{x_0} \frac{\delta j(x)}{j_{HoD}} \mathrm{dx} = \frac{0.026\mathrm{V}}{\mathrm{x_0 j_{HoD}}} \int_0^{x_0} \delta \mathrm{j(x)dx} \tag{2.15}$$

Since the total current is constant, i.e., the integral of the variation of current density *j(x)* is 0, $\delta V_{oc} = 0$.

In addition, the variance of light distribution is used to estimate the upper bound of the V_{oc}.in homogeneous illumination and that in heterogeneous illumination.

As Eq. 2.8 shown, Voc is represented as

$$V_{oc} = \frac{1}{n}\sum_{i=1}^{n} a\ln(bj_i + 1) = a\ln b + a\frac{1}{n}\sum_{i=1}^{n}\ln(P_i) \tag{2.16}$$

Where

$$P_i = j_i + \frac{1}{b} \tag{2.17}$$

Thus,

$$e^{V_{oc}} = e^{a\ln b}\left(\prod_{i=1}^{n} P_i\right)^{a/n} \tag{2.18}$$

In terms of a theorem [20] related to arithmetic mean geometric mean in Eq. 2.19, it gives a lower bound of Eq. 2.18 as Eq. 2.21 shown, which is related with the variance and average of $P_{i.}$

$$\prod_{i=1}^{n} x_i^{1/n} \geq \bar{x} - \frac{1}{2u}\sum_{i=1}^{n}\frac{1}{n}(x_i - \bar{x})^2 \tag{2.19}$$

$$x_i \in [u, v] \tag{2.20}$$

$$e^{V_{oc}} = e^{a\ln b}\left(\prod_{i=1}^{n} P_i\right)^{a/n} \geq e^{a\ln b}\left(\bar{P} - \frac{1}{2u}s\right)^{a} \tag{2.21}$$

Where

$$s = \frac{1}{n}\sum_{i=1}^{n} P_i^2 - \bar{P}^2 = \sigma \tag{2.22}$$

$$\bar{P} = \frac{1}{n}\sum_{i=1}^{n} P_i = \mu + \frac{1}{b} \tag{2.23}$$

$$\sigma = \frac{1}{n}\sum_{i=1}^{n} j_i^2 - \mu^2 \tag{2.24}$$

$$\mu = \frac{1}{n}\sum_{i=1}^{n} j_i \tag{2.25}$$

Then, the upper bound of V_{oc} is written as:

$$V_{oc} \geq a \ln b + a \ln\left(\bar{P} - \frac{1}{2u}s\right) = a \ln b + a \ln\left(\mu + \frac{1}{b} - \frac{1}{2u}\sigma\right) \tag{2.26}$$

The homogeneous illumination is represented as when distribution variance $\sigma = 0$ and the V_{oc} is written as Eq. 2.27 that corresponds with the V_{oc} formula of homogeneous case in manuscript.

$$V_{oc-\text{homo}} = a \ln b + a \ln\left(\mu + \frac{1}{b}\right) \tag{2.27}$$

Finally, the expected difference of homo-case and hetero-case to Voc is obtained by Eq. 2.28.

$$V_{oc-\text{homo}} - V_{oc-\text{heter}} \leq a \ln\left(\frac{2u(1 + b\mu)}{2u(1 + b\mu) - b\sigma}\right) \tag{2.28}$$

Since the average power μ of homo-case and hetero-case are the same, only the variance σ determines this upper bound in Eq. 2.28. In addition, for most polymer devices [21] and for most common distribution, $2u(1 + b\mu) >> b\sigma$, therefore V_{oc}-$V_{oc\text{-homo}} \sim 0$, i.e., homogeneous illumination and heterogeneous illumination show a similar impact on Voc. Beside the upper bound, we will give an accurate prediction in the next step research.

2.2 Heterogeneous Illumination

Now, we expand this principle to the architecture of a real device as shown in Fig. 2.4. First, it is imperative to know which variables are important and how they influence the optical power distributions, so we consider a specific geometry such as one single tube waveguide. As we have recently shown for fibers, which also holds true for tubes [7], the ratio of length to diameter can influence light absorption in the OCGOPVs, and as noted, this geometry also leads to a heterogeneous distribution of optical power in inner surface of this specific geometry. Further, there is a strong dependence of short-circuit current J_{sc}, on the incident angle at which the illumination is coupled. The optimum incident angle depends on this ratio of length to diameter of the structure Fig. 2.5 [7, 22].

Tube-based solar cell is an extended fiber device with a hemisphere bottom. They were fabricated on glass tubes with one end closed in a hemispherical cap (Chemglass, 1 mm I.D). The ITO films with a thickness of 100 nm were deposited on these *sub*strates by radio frequency magnetron sputtering (BOSCH) from an ITO target. (Depositing 100 nm ITO per 120° rotation, for three times.) The substrates

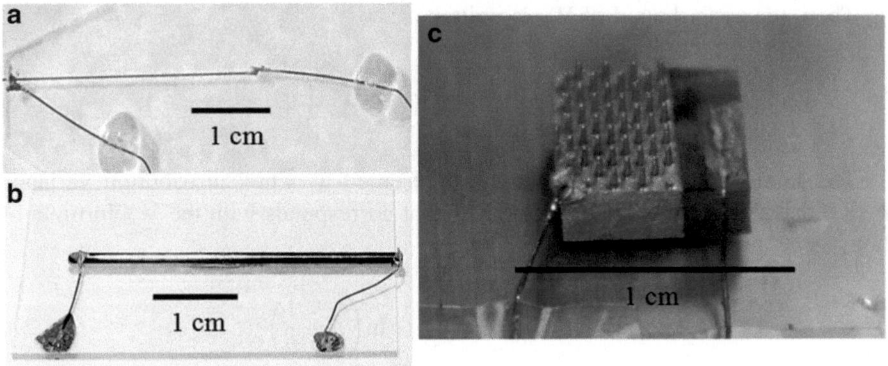

Fig. 2.4 (**a**) Fiber-based solar cell. (**b**) Tube-based solar cell. (**c**) Aligned plastic fiber cell with different diameter and length

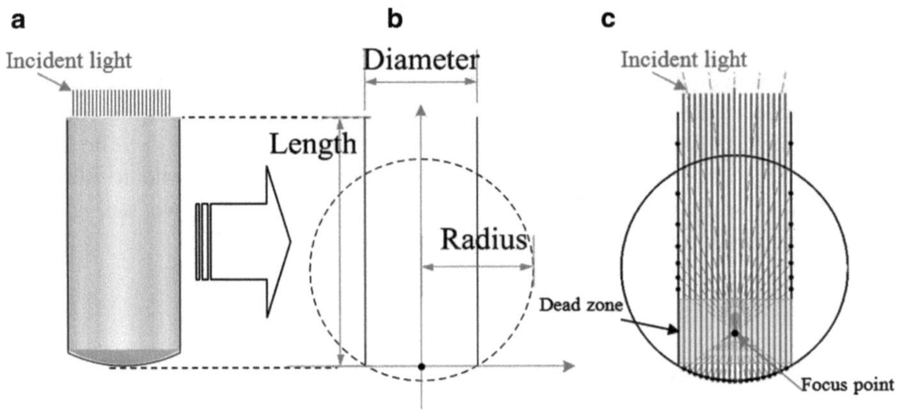

Fig. 2.5 (**a**) A OCGOPV with a given curvature cap at the bottom. (**b**) Light path in longitudinal section of (**a**). A beam of light (*red lines*, distance to y axis is x) enters into the tube (length h) and reflects at points (x1, y1) on bottom of curvature radius r, then reach the inner surface (*x*2, y2). (**c**) There is a "dead zone" near bottom, where no light shines. *Red lines* (*solid lines*) are incident light paths; and *gray lines* (*dashed lines*) are reflected light paths

were then exposed to ozone for 90 min (rotating the tubes three times every 30 min). Subsequently, the buffer layer and absorber layer were added by dip coating. Poly (3,4-ethylenedioxythiophene)-poly(styrenesulfonate) (PEDOT:PSS, Clevios P, the film thickness is ~40 nm), and poly 3-hexyl thiophene:phenyl-C_{61}-butyric acid methyl ester (P3HT:PCBM = 1:0.8 in WT in chlorobenzene, of 15 mg/ml P3HT for planar cell and 5 mg for tube devices) were deposited on the tube substrate. Finally, Al electrodes were deposited via thermal evaporation at the pressure of 10^{-6} torr. Similarly, planar devices were fabricated using spin coating, for comparison. Both devices were lightly annealed (~100°C for 5 min).

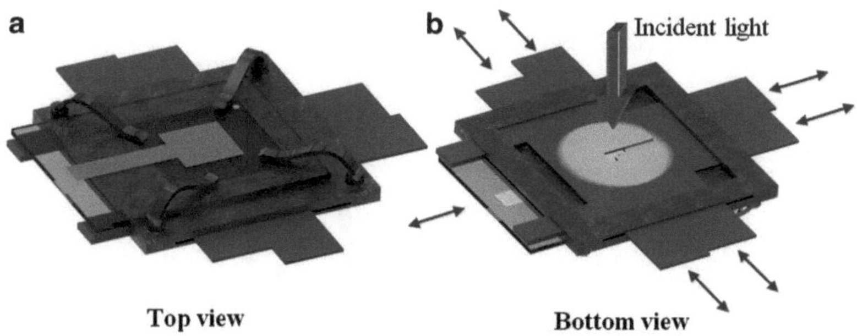

Fig. 2.6 A tool for measuring heterogeneous distribution of solar light in a subunit of a planar device. Solar light comes from the bottom (*Red arrow* (*light arrows*) in (**b**) shows the direction of solar light), and goes through the aperture hole for which the size could be adjusted by four panels. (*Blue arrows* (*dark arrows*) represent the adjustable directions of slices in (**a**))

Table 2.1 Sample fabrication details for heterogeneous illuminations

Sample	Architecture	Fabrication method
A	Substrate/ITO/PEDOT/P3HT:PCBM/Al	**Spin coating**
B	Tube/ITO/PEDOT/P3HT:PCBM/Al	**Dip coating**
C	Substrate/ITO/PEDOT/P3HT:PCBM/Al	**Dip coating**

For both device types, the illumination intensity was varied to determine the performance as a function of illumination power. The case of heterogeneous illumination was simulated for the planar device by changing the area of illumination by a special mask for a given luminous power. The architecture of this tool is shown in Fig. 2.6. The active area is 0.5 cm^2. Current voltage characteristics were collected using Keithley 236 source-measurement unit and an AM1.5 g standard Newport # 96000 Solar Simulator. The output power intensity is adjusted from 0.5 to 150 mW/cm^2.

To test the above model three different types of organic photovoltaic cells were built. All are based on the P3HT:PCBM bulk heterojunction as described. Architectures and corresponding fabrication methods are shown in Table 2.1.

As shown above, both homogeneous and heterogeneous cases have the same functional dependence with voltage. If this is true, it will provide a simple method to study the global distribution of optical power in the OCGOPVs. First, we use an "ideal" planar device (sample A) to test the two illumination cases, shown in Fig. 2.7. As expected, both HeOI and HoOI show the same overall impact on the observed Voc as average input (at the front of the device) light intensity is lowered. Specifically, the Voc drops sharply when input optical intensity is lower than ~ 10mW/cm^2, which is also described by Eq. 2.3 above. In other words, the variation of the voltage is the same for every case where only a small subunit is illuminated, or part of the whole cell is illuminated. The gray curve is fitted by Eq. 2.3 in which all the parameters of devices are extracted from IV curve in terms of an iterative method [21].

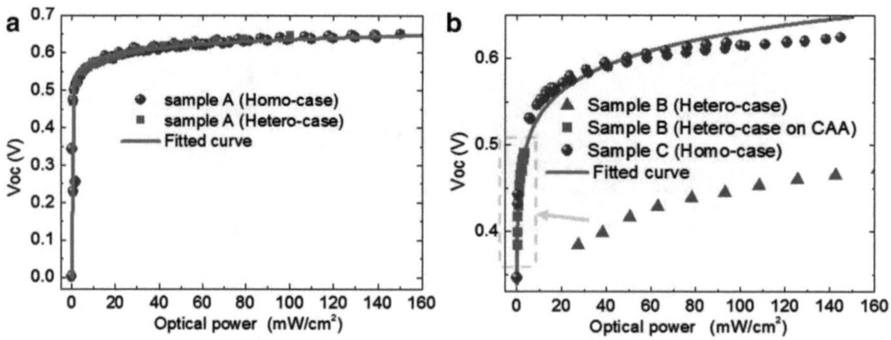

Fig. 2.7 (**a**) The Voc versus optical intensity of sample A, which is illuminated by simulated heterogeneous variation (*red square*) and homogeneous variation (*blue sphere*) respectively. The Voc versus I curves are fitted to Eq. 2.3 and shown as the *grey line*. (**b**) The Voc versus optical intensity for dip coating devices. The curve with "on CAA" means the optical power of incident optical flux divided CAA [5]. Here, the *red square* and *green triangle* are the same illumination data divided by IAA and CAA respectively

Then, we apply this principle to the tube-based cell (sample B), which was prepared by a dip coating method. In this case, the usual incident light at the front face of the OCGOPV (sample B) is homogeneous, but it will be distributed throughout the volume heterogeneously, on the inner surface as in Fig. 2.4d. In Fig. 2.7b, we plot a planar OPV (sample C, fabricated using the same dip coating procedure) together with the tube-based device. Because OCGOPVs have a very large CAA, i.e., the average optical intensity on the inner surface (CAA) is very low, the Voc of the tube OPV tends to be a lower voltage than planar devices as shown in green. To determine the average optical power inside the tube, the total incident flux was divided by the total internal area of the device. This effective heterogeneous intensity (red square, B) was used to normalize the Voc to intensity. This normalized Voc shows the same functional behavior in the OCGOPV as homogeneous illumination does in the control device (Blue sphere, C).

To fully understand how this may modify the overall performance of the OCGOPV (the efficiency), we have also measured the filling factor and current collection for the planar device and compared it to the OCGOPV as a function of illumination intensity. Figure 2.8a shows the FF for the spin cast device (sample A, our "ideal" device), the OCGOPV (sample B) and the dip coated planar device (sample C). For the planar device fabricated with spin coating, there exists a peak in FF near 10 W/m^2. As is typical with such OPVs, the FF varies with the quality of the thin films, but it can be rather high (near 0.7). Generally, FF of OCGOPV is also limited by the film quality. Since dip coating was used to fabricate tube-based cell, resulting in uneven films, the FF for both the dip-coated tube structures and planar structures drop. In the case of the tube device, it is approximately reduced by 0.25. The normalized FF of OCGOPV also shows the same functional behavior as homogeneous illumination does in dip-coated planar devices (Pink line, C). In terms of Eq. 2.1, J-V is simulated to find the maximum obtainable output power

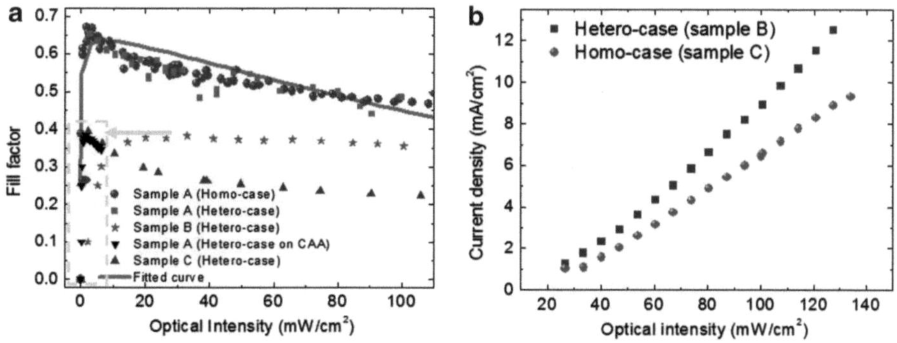

Fig. 2.8 (**a**) FF versus optical intensity of sample A, sample B and sample C. The *green stars* and the *black down-triangles* (in *yellow rectangle*) are the same input flux data divided by IAA and CAA respectively. The *gray line* is the fitted curve by Eq. 2.1. (**b**) Jsc versus optical intensity (IAA) of planar and OCGOPV device with architecture ITO/PEDOT/P3HT:PCBM/Al

P_m and corresponding J_m and V_m, then the FFs of different Jph are obtained as gray line in Fig. 2.8a to compare with the experiment.

On the other hand, Fig. 2.8b illustrates that the OCGOPV can transfer more light flux to current than the conventional OPV. This has been previously reported in several publications [22–24]. Moreover, this advantage is enhanced with increasing optical illumination intensity at the entrance aperture. This suggests that OCGOPVs will perform well with high illumination intensity applications such as concentrators [25–27].

We can now examine the overall result of these effects on efficiency. The efficiency as a function of incident optical illumination intensity is shown in Fig. 2.9a. For the planar cell (sample A, Blue spheres), the optimum optical intensity of the highest efficiency is near 10 mA/cm^2, which is far less than the AM1.5 g standard solar light (100 mW/cm^2) allowed for by P3HT:PCBM. That indicates the highest performance requires reducing the optical intensity. However, the tube-based cell (sample B) exhibits a monotonically increasing efficiency with optical intensity at the input aperture. This is because at the lowest values of IAA, the optical intensity within the device volume is very low. When the efficiency of the OCGOPV is scaled with the average internal illumination intensity given by CAA (shown in the yellow rectangle), the curve's shape is similar to that of the planar cell of both spin cast and dip coat at lower optical intensity but performs better than that of planar cell of same dip coating procedure (Upper angle, C). This also suggests the functional equivalence of HoOI and HeOI in OPVs, generally.

Using this "optical intensity effect," we can predict the optimum efficiency of OCGOPVs, as a function of the geometry: ratios of the length/diameter of OCGOPV together with the ratios of bottom curvature radius to diameter (radius/diameter). Since the two fabrication procedures, dip coating film and spin coating, yield different film morphologies, we must also consider this. In Fig. 2.9b, we again use the tube-based cell as a typical case of OCGOPVs. The efficiency drops with the

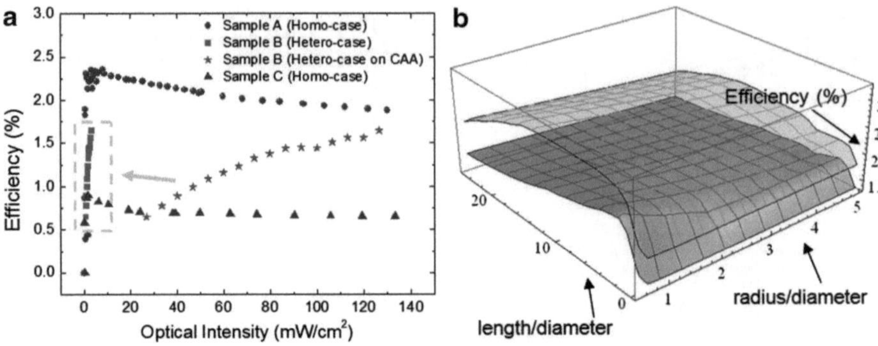

Fig. 2.9 (**a**). Efficiency versus optical intensity of sample A, sample B and sample C. The *green stars* and the *red squares* (in *yellow rectangle*) are the same input flux data divided by IAA and CAA respectively. (**b**) The simulation of efficiency versus the ratio length/diameter of tube, and the ratio of bottom curvature radius and tube diameter, is in the condition of normal incident light and for two fabrications of spin coating (*Green surface*) and dip coating (*Orange surface*). This simulation is based on an optical path method and a transfer matrix mentioned before [7, 17], and also use the experiment data from the *left figure* (**a**)

increase of the ratio of radius of bottom/diameter of tube, because of the reflection loss from the bottom reflector (see in Fig. 2.5c) which we have termed "the dead zone" (when this ratio $> 1/\sqrt{2}$). Moreover, the peak in the dimension of length/diameter is the combined result of current increase and voltage decrease. Longer OCGOPVs can absorb more light energy to generate a higher current density [7], but it has a negative impact on voltage. This phenomenon is simulated for both film cases. The peaks are all with relatively low optical intensity ($<<100$ mW/cm^2) on the inner surface. Thus, the optimum design for OCGOPVs should be of length/diameter that is in the range of 1–5 and the curvature radius/diameter is between 0.5 and 0.7 for normal incidence, which the accurate ratio predictions depend on the film quality, thickness and polymer type. In Fig. 2.9b, the optimum design of tube-based geometries using P3HT:PCBM is located at length/diameter 2 and radius/diameter 0.7. This could explain why in the past, research showed that longer fiber-devices were used to reach high photocurrent but led to a lower voltage [5, 28].

In sum, this section has examined the overall effect of 3D architectures by heterogeneous illumination. Specifically, we have provided an equivalent circuit equation of OCGOPVs by revising the Shockley equation to apply to the case of optical heterogeneity within the structure. Using this approach, we have shown a functional equivalence of internal HoOI and HeOI for OPVs generally. The overall effect of optical heterogeneity on performance parameters in our model, 3D, fiber-based structure is that the Voc is lowered generally as the ratio of length to diameter becomes larger. This corresponds to a large internal area of the device. Jsc is seen to increase substantially for large internal device areas whereas the FF is less sensitive overall. We have correlated our model calculations to experimentally obtained values in a fiber device using P3HT:PCBM absorbers. Finally, it was found that there is an optimum design to the highest efficiency for tube-based OCGOPV, which corresponds to a length/diameter in the range of 1–5, and an optimum radius of bottom/diameter of tube lower than 0.7. These results suggest that realization of truly high performance devices utilizing polymers on 3D architectures will require new polymer systems capable of maintaining higher voltages under lower light conditions.

2.3 Dependence of Light Illumination Intensity

Fiber-based architecture, which is designed to confine the incident light in the fiber where it bounces around until it is absorbed almost completely, has been proven to not only lead to better optics performance but also a relatively thinner absorber layer reducing recombination rate. In this section, with theoretical simulation, we further demonstrate that, under high intensity of sun light (>1,000 suns), the fiber-based structure can dramatically enhance the efficiencies by 52.5% of planar solar cell while extending the prime working light intensity range from dozens of suns to thousands. This result shows that fiber architecture has great application potential in high concentration photovoltaic (HCPV) system.

For years, concentration photovoltaic (PVC) has attracted researchers for both very high conversion efficiencies [29–32], and dramatic reduction in the module cost for photovoltaic electricity generation by decreasing the amount of semiconductor materials needed. However, the development has been constrained by the lack of ability resisting the high intensity of light while maintaining a good performance. As we know, the efficiency will be minimized by high light intensity because efficiency equals to the product of open circuit voltage (Voc), short current density (Jsc), and fill factor (FF), which are mainly determined by the band gap and temperature. Although light intensity is beneficial to Voc [33], in contrast, high temperature resulting from very high light intensity will reduce Voc, and raise the recombination rate causing lower Jsc. Furthermore, intense current in the equivalent circuit of a solar cell would further provide negative influence on FF and a module's lifetime. Consequently, all those facts minify the efficiency at high light intensity. However, a good PVC module should be capable of performing well at very high illumination intensity (>1,000 suns) and also have a large working range considering sun light intensity is volatile for different time or weather conditions. To solve this conflict, advantages of fiber-based solar cells should be considered. This optical confinement geometry has been developing in recent years [6, 7, 24, 34, 35], mostly using fiber and tube to confine light in its cavity and fully absorb light in the range of absorption. Our former work argued that the fiber-based solar cell (FBSC) performs better at higher illumination [36] due to fiber's large aspect ratio. The current active area (the wall) is much larger than the light incident area (the ends) which means the light density (incident light/active area) is low enough to avoid all the negative effects discussed above. In this work, we further examined the performance at very high illumination (>1,000suns) and the universality of FBSC on different materials Fig. 2.10.

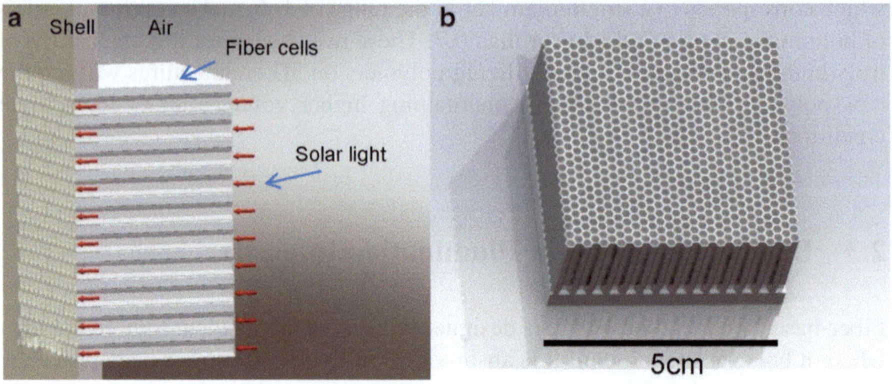

Fig. 2.10 (**a**) A cross section of fiber-based photovoltaic system. (**b**) An integrated unit of fiber-based solar cells. *Red arrows* represent incident light

In the past, the performances of planar solar cells based on several types of materials have been reported, including silicon [37], GaInP/GaInAs/Ge [38], CIGS [39], CZTS [40], polymer [41] and small molecular [42]. Besides, at certain incident light illumination, the performances were simulated by the following methods. First, the light distribution along the fiber cavity was obtained by a ray tracing method in Ref. [7], and the optical field in active layer is calculated in terms of Ref. [43]. By dividing fiber cell into many infinitesimal planar cells, its equivalent circuit was obtained [36]. Its photocurrent (J_{ph}), saturated current (J_s), ideal factor (n_D), and resistances of device could be extracted from the current-voltage (IV) curves with Lambert W-function [44] which is used to estimate the parameters in Poisson and continuity equations [45]. The circuit response at certain light intensity input was then simulated by finding the stable solution using Gummel's method [46]. Figure 2.11 shows the efficiencies prediction versus incident light intensity of these six materials, for both fiber cells and planar cells. All kinds of devices, regardless of being fiber-based or planar-based, have a max efficiency at a certain optimum light intensity, which are usually 100 suns for planar cells. For fiber cells, not only the prime working light intensities are increased to 593 ~ 4,830 suns for different materials, but also the max efficiencies were improved. Table 2.2 lists the best efficiencies (Eff) and corresponding light intensities (I). Remarkably, the GaInP/GaInAs/Ge triple-junction cell's efficiency reaches 43% on a fiber-based cell, which is much higher than the planar one, and a similar efficiency (40%) has been achieved at 240 suns in practice [47]. In addition, FBSC keeps 80% of the highest efficiency in a range of 7,741 suns at very high light intensity. Compared to only 35.5 suns range of planar cells at a relatively low light intensity, FBSA has much better compatibility, especially for concentrating solar power system. Furthermore, it is surprising that even organic cells like polymer and small molecular can both reach 10% efficiency, which is a general critical value to realize industrial production.

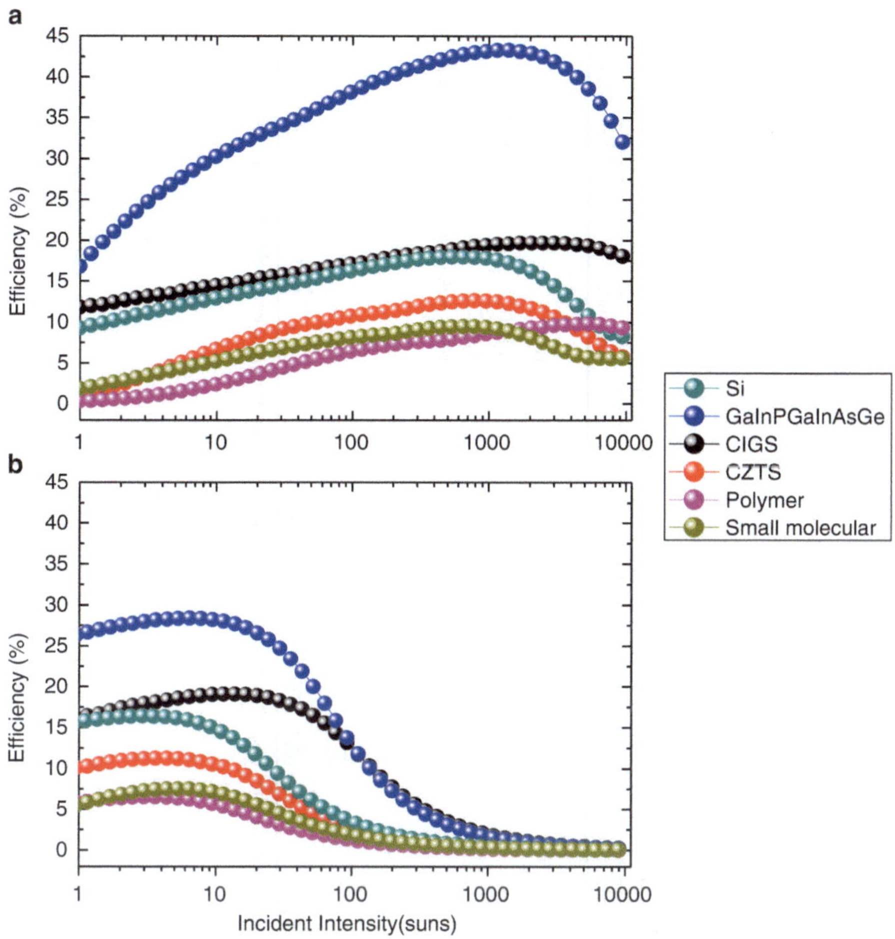

Fig. 2.11 Efficiencies versus incident intensity for fiber-based solar cells in (**a**) and planar solar cells in (**b**)

Figure 2.12 further shows the efficiency dependence on incident angle and aspect ratio for fiber cells at 1,000 suns light intensity. It generally reveals that the optimum incident angles are lower than 40°, and become smaller along with the increase of aspect ratio, because the total light flux passing through the fiber is proportional to the cosine function of incident angle, such that more energy will be wasted before entering the fiber's cavity for bigger angles. In addition, although longer fiber is beneficial to the absorption, meanwhile, the larger CCA leads to sharp Voc dropping off [36], therefore giving the aspect ratios an optimum value. This provides a concept of designing the architecture of fiber-based solar used in a concentrator system.

Table 2.2 Max efficiencies (Eff) and corresponding light intensity (I) for six materials based on fiber cells and planar cells

	Si	GaInP/GaInAs/Ge	CIGS	CZTS	Polymer	Small molecular
Fiber cells						
Eff (%)	18.0	43.3	19.7	12.6	9.8	9.5
I (suns)	593	1,272	2,479	790	4,830	653
R_{work} (suns)	2,970	7,741	9,381	3,239	8,922	2,193
	(28 ~ 2,999)	(37 ~ 7,779)	(31 ~ 9,412)	(60 ~ 3,299)	(490 ~ 9,412)	(60 ~ 2,257)
Planar cells						
Eff (%)	16.4	28.4	19.3	11.3	7.5	6.5
I (suns)	3	6	13	4	5	3
R_{work} (suns)	14 .4	35.5	76.5	16.2	15.4	10.9
	(0.2 ~ 14.6)	(0.2 ~ 35.7)	(0.2 ~ 76.7)	(0.5 ~ 16.7)	(1.3 ~ 16.7)	(0.5 ~ 11.4)

Note: R_{work} represents working range that device has efficiency higher than 80% max efficiency, and with the start and end light intensity in brackets

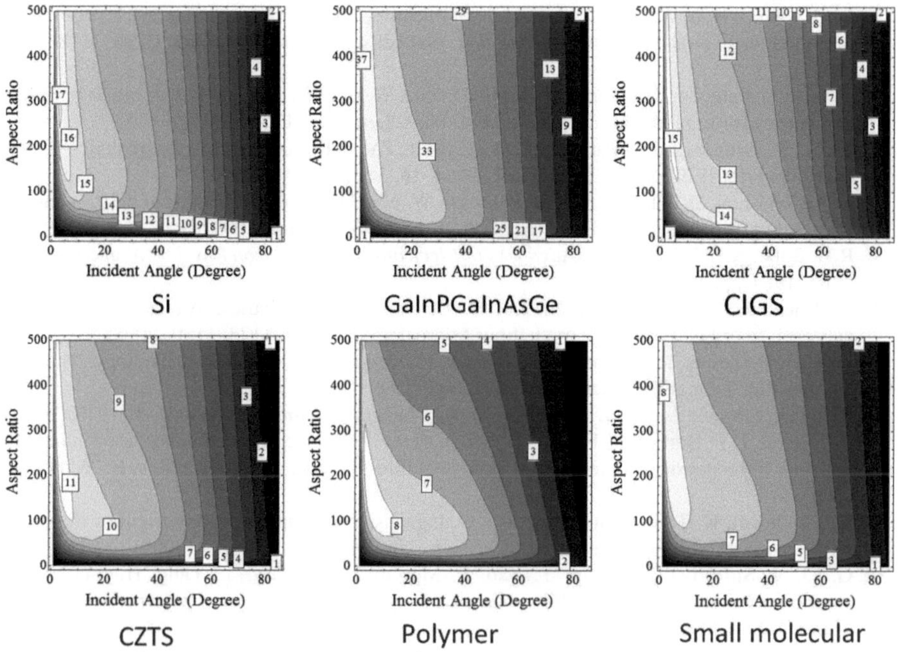

Fig. 2.12 Efficiencies (%) versus incident angle and aspect ratio for fiber-based solar cells based Si, GaInPGaInAsGe, CIGS, CZTS, polymer and small molecular

In sum, fiber-based solar cell has an excellent performance at very high light intensity by increasing the max efficiency and prime working light intensity as well as extending the working range hundreds of times of the planar cells. Also, all those advantages are proved to be valid for six types of solar cells, especially for GaInP/GaInAs/Ge solar cell, which is mostly used in HCPV system nowadays. Therefore, this architecture may provide a brand new method and different thought to clearing the obstacles facing the development of a CPV system.

References

1. Y. Li, M. Wang, H. Huang, W. Nie, Q. Li, E.D. Peterson, R. Coffin, G. Fang, D.L. Carroll, Influence on open-circuit voltage by optical heterogeneity in three-dimensional organic photovoltaics. Phys. Rev. B. **84**(8), 085206
2. R. Hausermann, E. Knapp, M. Moos, N.A. Reinke, T. Flatz, B. Ruhstaller, Coupled optoelectronic simulation of organic bulk-heterojunction solar cells: Parameter extraction and sensitivity analysis. J. Appl. Phys. **106**, 104507 (2009)
3. D.W. Sievers, V. Shrotriya, Y. Yang, Modeling optical effects and thickness dependent current in polymer bulk-heterojunction solar cells. J. Appl. Phys. **100**, 114509 (2006)

4. Y.M. Nam, J. Huh, W.H. Jo, Optimization of thickness and morphology of active layer for high performance of bulk-heterojunction organic solar cells. Sol. Energ. Mat.Sol. C. **94**, 1118–1124 (2010)
5. Y. Li, E.D. Peterson, H. Huang, M. Wang, D. Xue, W. Nie, W. Zhou, D.L. Carroll, Tube-based geometries for organic photovoltaics. Appl. Phys. Lett. **96**, 243505 (2010)
6. Y. Li, W. Nie, J. Liu, A. Partridge, D.L. Carroll, The optics of organic photovoltaics: Fiber-based devices. IEEE J. Sel. Top. Quant. Elect. **16**, 1827–1837 (2010)
7. Y. Li, W. Zhou, D. Xue, J. Liu, E.D. Peterson, W. Nie, D.L. Carroll, Origins of performance in fiber-based organic photovoltaics. Appl. Phys. Lett. **95**, 203503 (2009)
8. R.H. Bube, A.L. Fahrenbruch, *Advances in Electronics and Electron Physics* (Academic, New York, 1981), p. 163
9. A. Cheknane, H.S. Hilal, F. Djeffal, B. Benyoucef, J.P. Charles, An equivalent circuit approach to organic solar cell modelling. Microelectron. J. **39**, 1173–1180 (2008)
10. A. Shah, P. Torres, R. Tscharner, N. Wyrsch, H. Keppner, Photovoltaic technology: The case for thin-film solar cells. Science **285**, 692–698 (1999)
11. A.L. Fahrenbruch, J. Aranovich, *Solar Energy Conversion* (Springer, New York, 1979), p. 257
12. B.P. Rand, D.P. Burk, S.R. Forrest, Offset energies at organic semiconductor heterojunctions and their influence on the open-circuit voltage of thin-film solar cells. Phys. Rev. B. **75**, 115327 (2007)
13. K. Vandewal, K. Tvingstedt, A. Gadisa, O. Inganas, J.V. Manca, On the origin of the open-circuit voltage of polymer-fullerene solar cells. Nat. Mater. **8**, 904–909 (2009)
14. G. Li, V. Shrotriya, J. Huang, Y. Yao, T. Moriarty, K. Emery, Y. Yang, High-efficiency solution processable polymer photovoltaic cells by self-organization of polymer blends. Nat. Mater. **4**, 864–868 (2005)
15. Y. Li, Y. Zou, Conjugated polymer photovoltaic materials with broad absorption band and high charge carrier mobility. Adv. Mater. **20**, 2952–2958 (2008)
16. A. Pivrikas, G. Juscaronka, A.J. Mozer, M. Scharber, K. Arlauskas, N.S. Sariciftci, H. Stubb, R.O. Sterbacka, Bimolecular recombination coefficient as a sensitive testing parameter for low-mobility solar-cell materials. Phys. Rev. Lett. **94**, 176806 (2005)
17. L.A.A. Pettersson, L.S. Roman, O. Inganas, Modeling photocurrent action spectra of photovoltaic devices based on organic thin films. J. Appl. Phys. **86**, 487–496 (1999)
18. F. Monestier, J.-J. Simon, P. Torchio, L. Escoubas, F. Flory, S. Bailly, R. de Bettignies, S. Guillerez, C. Defranoux, Modeling the short-circuit current density of polymer solar cells based on P3HT:PCBM blend. Sol. Energ. Mat. Sol. C. **91**, 405–410 (2007)
19. B.G. Wei, Study of parallel DC power. Tech. Phys. Teach. **16**, 17–18 (2008)
20. D.I. Cartwright, M.J. Field, Refinement of arithmetic mean geometric mean inequality. P. Am. Math. Soc. **71**, 36–38 (1978)
21. K.I. Ishibashi, Y. Kimura, M. Niwano, An extensively valid and stable method for derivation of all parameters of a solar cell from a single current–voltage characteristic. J. Appl. Phys. **103**, 094507 (2008)
22. J.W. Liu, M.A.G. Namboothiry, D.L. Carroll, Optical geometries for fiber-based organic photovoltaics. Appl. Phys. Lett. **90**, 133515 (2007)
23. M.R. Lee, R.D. Eckert, K. Forberich, G. Dennler, C.J. Brabec, R.A. Gaudiana, Solar power wires based on organic photovoltaic materials. Science **324**, 232–235 (2009)
24. B. O'Connor, K.P. Pipe, M. Shtein, Fiber based organic photovoltaic devices. Appl. Phys. Lett. **92**, 193306 (2008)
25. M.J. Currie, J.K. Mapel, T.D. Heidel, S. Goffri, M.A. Baldo, High-efficiency organic solar concentrators for photovoltaics. Science **321**, 226–228 (2008)
26. C. Dominguez, I. Anton, G. Sala, Solar simulator for concentrator photovoltaic systems. Opt. Express **16**, 14894–14901 (2008)
27. K. Omer, H. Baruch, A.K. Eugene, M.G. Jeffrey, High-flux characterization of ultrasmall multijunction concentrator solar cells. Appl. Phys. Lett. **91**, 064101 (2007)

28. H. Huang, Y. Li, M. Wang, W. Nie, W. Zhou, E.D. Peterson, J. Liu, G. Fang, D.L. Carroll, Photovoltaic-thermal solar energy collectors based on optical tubes. Sol. Energ. **85**, 450–454 (2011)
29. L.W. James, R.L. Moon, Gaas concentrator solar cell. Appl. Phys. Lett. **26**, 467–470 (1975)
30. R. Sinton, Y. Kwark, R.M. Swanson, 23-percent efficient Si point contact concentrator solar-cell. IEEE T. Electron. Dev. **32**, 2553–2554 (1985)
31. M.A. Green, A. Ho-Baillie, Forty three per cent composite split-spectrum concentrator solar cell efficiency. Prog. Photovoltaics **18**, 42–47 (2010)
32. J.S. Ward, K. Ramanathan, F.S. Hasoon, T.J. Coutts, J. Keane, M.A. Contreras, T. Moriarty, R. Noufi, A 21.5% efficient Cu(In,Ga)Se-2 thin-film concentrator solar cell. Prog. Photovoltaics **10**, 41–46 (2002)
33. S.R. Cowan, A. Roy, A.J. Heeger, Recombination in polymer-fullerene bulk heterojunction solar cells. Phys. Rev. B. **82**, 245207 (2010)
34. B. Weintraub, Y.G. Wei, Z.L. Wang, Optical fiber/nanowire hybrid structures for efficient three-dimensional dye-sensitized solar cells. Angew. Chem. Int. Edit. **48**, 8981–8985 (2009)
35. A. Gagliardi, M.A.D. Maur, A. Di Carlo, Theoretical investigation of a dye solar cell wrapped around an optical fiber. IEEE J. Quantum. Elect. **47**, 1214–1221 (2011)
36. Y. Li, M. Wang, H. Huang, W. Nie, Q. Li, E.D. Peterson, R. Coffin, G. Fang, D.L. Carroll, Influence on open-circuit voltage by optical heterogeneity in three-dimensional organic photovoltaics. Phys. Rev. B. **84**, 085206 (2011)
37. S.B. Ghozati, A.U. Ebong, C.B. Honsberg, S.R. Wenham, Improved fill-factor for the double-sided buried-contact bifacial silicon solar cell. Sol. Energ. Mat. Sol. C. **51**, 121–128 (1998)
38. S. Roensch, R. Hoheisel, F. Dimroth, A.W. Bett, Subcell I-V characteristic analysis of GaInP/GaInAs/Ge solar cells using electroluminescence measurements. Appl. Phys. Lett. **98**, 251113 (2011)
39. NREL, Manufacturing-friendly advance seen in CIGS solar cell processing, NREL Highlights, NREL/FS-5200-51412 (2011)
40. T.K. Todorov, K.B. Reuter, D.B. Mitzi, High-efficiency solar cell with earth-abundant liquid-processed absorber. Adv. Mater. **22**, E156 (2010)
41. W. Nie, C.M. MacNeill, Y. Li, R.E. Noftle, D.L. Carroll, R.C. Coffin, A soluble high molecular weight copolymer of benzo[1,2-b:4,5-b']dithiophene and benzoxadiazole for efficient organic photovoltaics. Macromol. Rapid. Comm. (2011) n/a-n/a
42. M. Riede, C. Uhrich, J. Widmer, R. Timmreck, D. Wynands, G. Schwartz, W.M. Gnehr, D. Hildebrandt, A. Weiss, J. Hwang, S. Sundarraj, P. Erk, M. Pfeiffer, K. Leo, Efficient organic tandem solar cells based on small molecules. Adv. Funct. Mater. **21**, 3019–3028 (2011)
43. Y. Li, H. Huang, M. Wang, W. Nie, W. Huang, G. Fang, D.L. Carroll, Spectral response of fiber-based Photovoltaics. Sol. Energ. Mat. Sol. C. (2011 in press)
44. A. Jain, A. Kapoor, A new method to determine the diode ideality factor of real solar cell using Lambert W-function. Sol. Energ. Mat. Sol. C. **85**, 391–396 (2005)
45. L.J.A. Koster, E.C.P. Smits, V.D. Mihailetchi, P.W.M. Blom, Device model for the operation of polymer/fullerene bulk heterojunction solar cells. Phys. Rev. B. **72**, 085205 (2005)
46. H.K. Gummel, A self-consistent iterative scheme for one-dimensional steady state transistor calculations. IEEE T. Electron. Dev. **11**, 455–465 (1964)
47. R.R. King, D.C. Law, K.M. Edmondson, C.M. Fetzer, G.S. Kinsey, H. Yoon, R.A. Sherif, N.H. Karam, 40% efficient metamorphic GaInP/GaInAs/Ge multijunction solar cells. Appl. Phys. Lett. **90**(18), 183516 (2007), AIP

Chapter 3
Optical Path in Cavity

3.1 Optical Path Model

Using ray tracing and optical path iteration, this section present a mathematical model for light transmission, absorption, and loss in fiber-based organic photovoltaics (Reprinted with permission from [1], Copyright 2009, American Institute of Physics). Simulations based on this model give an optimum incident angle, position and an optimum aspect ratio in terms of other parameters of the fiber photocell. Further, the predicted relationship between current generation and incident angle correlates well with experimental data for a given fiber length. Finally, we show that experimental results on current generation versus fiber diameter are well reproduced by our simulation for fiber diameters of the order of the wavelength of incident light.

The fiber photovoltaic concept addresses the shortcoming of planar solar cell by wrapping the photovoltaic device around an optical fiber [2]. Light is then incident on the fiber end and transmitted down the fiber allowing for multiple passes through the active layer forming what is essentially an "optical can." Modeling of such a structure using ray optics allows us to understand the optimum performance and response characteristics of such cells. Earlier work [3] employing such modeling is incomplete in that it ignores multiple internal reflections. In this section, we explore how all the processes involved in light transmission, absorption and loss, can be modeled by following a combined iterative approach. The result predicts a surprising increase in overall optimum performance from an organic cell in the fiber photovoltaic configuration, which is different from that predicted in earlier work.

As shown in Fig. 3.1, thc fiber has four layers deposited on it (ITO, PEDOT, P3HT, Al), with refractive indexes of 1.45, 1.9, 1.5, and 1.6, respectively.

Y. Li, *Three Dimensional Solar Cells Based on Optical Confinement Geometries*, Springer Theses, DOI 10.1007/978-1-4614-5699-5_3,

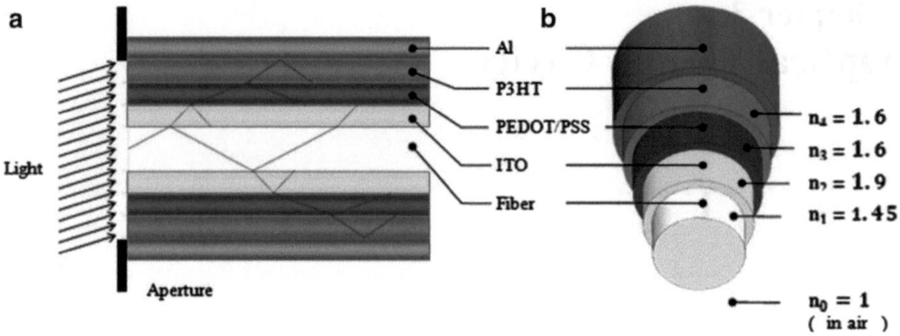

Fig. 3.1 Schematic of the fiber solar cell architecture and light illumination, ray diagram of light propagation

3.1.1 *Incidence and Transmission*

As shown in Fig. 3.2a, the incident angle is divided into a transverse angle θ_i and tangential angle φ_i (we assume an initial value of φ_1 of 90°). In terms of Snell's Law, the transverse angles θ_i are given by

$$n_i \sin\theta_i = n_t \sin\theta_t \tag{3.1}$$

Then, we put each angle into Eq. 3.1, so we obtain:

$$n_0 \sin\theta_0 = n_1 \sin\theta_1, \theta_2 = \frac{\pi}{2} - \theta_1 \tag{3.2a}$$

$$n_1 \sin\theta_2 = n_2 \sin\theta_3, \theta_4 = \frac{\pi}{2} - \theta_3 \tag{3.2b}$$

$$n_2 \sin\theta_3 = n_3 \sin\theta_5, \theta_6 = \frac{\pi}{2} - \theta_5 \tag{3.2c}$$

$$n_3 \sin\theta_5 = n_4 \sin\theta_7, \theta_8 = \frac{\pi}{2} - \theta_7 \tag{3.2d}$$

Likewise, in Fig. 3.3c, the equations for each angle and refractive index are the following.

$$\varphi_1 = \frac{\pi}{2}(arbitrary),\ \varphi_2 = \frac{\pi}{2} - \tan^{-1}\frac{y_1}{x_1},\ \varphi_3 = \sin^{-1}\left[\frac{n_1}{n_2}\sin(\varphi_2)\right] \tag{3.3}$$

In terms of the Cosine Theorem in triangle OP_1P_2,

$$P_1P_2^2 + OP_1^2 - 2P_1P_2OP_1\cos(\pi - \varphi_3) = OP_2^2 \tag{3.4}$$

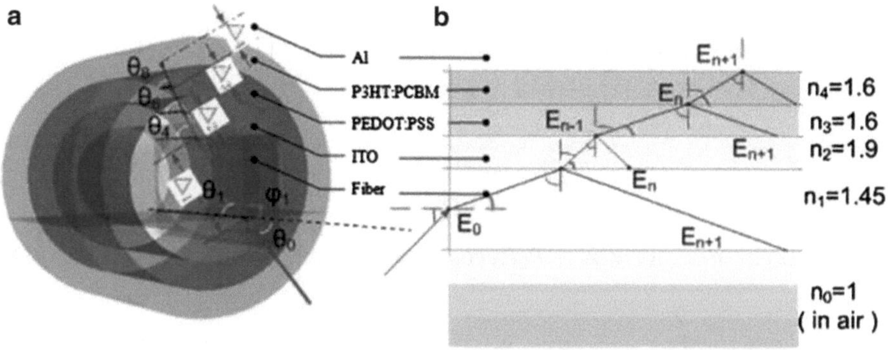

Fig. 3.2 (**a**) Cross section of light path in fiber. Light enters the front face and is transmitted into each layer with transverse angle θ_i, tangential angle φ_i, and optical path Δ_i, which could be obtained by Snell's Law and Geometrical optics (**b**) Longitudinal section of light path in fiber. The refractive indexes of P3HT:PCBM, PEDOT, ITO and fiber are 1.6, 1.6, 1.9, 1.45 respectively. E_i is remaining energy on certain refracted point, which comes from the previous refracted point E_{i-1} (previous remaining energy)

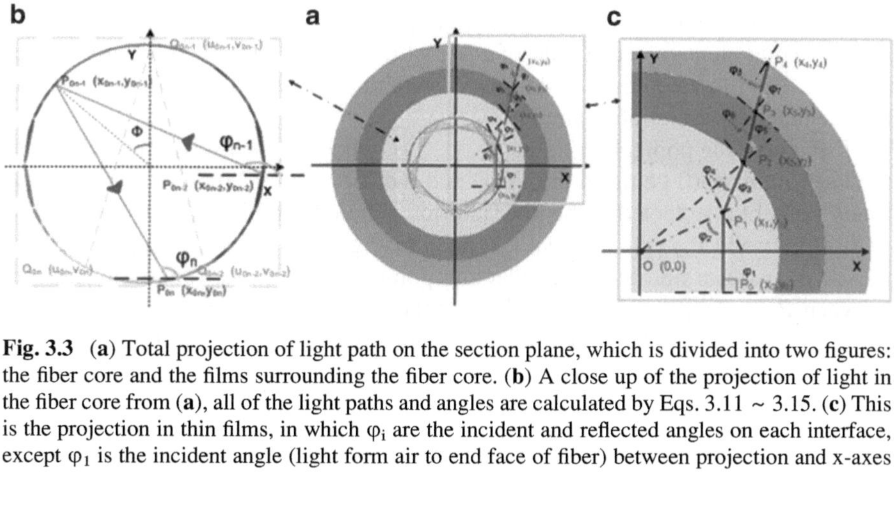

Fig. 3.3 (**a**) Total projection of light path on the section plane, which is divided into two figures: the fiber core and the films surrounding the fiber core. (**b**) A close up of the projection of light in the fiber core from (**a**), all of the light paths and angles are calculated by Eqs. 3.11 ~ 3.15. (**c**) This is the projection in thin films, in which φ_i are the incident and reflected angles on each interface, except φ_1 is the incident angle (light form air to end face of fiber) between projection and x-axes

We can obtain the quantity of P_1P_2 and the coordinate of P_2 (x_2, y_2),

$$P_1P_2 = -OP_1 \cos\varphi_3 + \sqrt{OP_2^2 - OP_1^2 + OP_1^2\cos^2\varphi_3} \tag{3.5}$$

$$P_2: \ x_2 = x_1 + P_1P_2 \cos\varphi_3, y_2 = y_1 + P_1P_2 \sin\varphi_3 \tag{3.6}$$

The incident angles, refractive angles, and coordinate values of transmission point for each layer can be obtained as the following:

$$\varphi_4 = \sin^{-1}\left(\frac{y_2 - y_1}{\sqrt{(x_2 - x_1)^2 + (y_2 - y_1)^2}}\right) - \sin^{-1}\left(\frac{y_2}{\sqrt{x_2^2 + y_2^2}}\right) \tag{3.7a}$$

$$\varphi_5 = \sin^{-1}\left[\frac{n_2}{n_3}\sin(\varphi_4)\right] \tag{3.7b}$$

$$P_3: \quad x_3 = x_2 + P_2P_3\cos\varphi_5, \; y_3 = y_2 + P_2P_3\sin\varphi_5 \tag{3.8}$$

$$\varphi_6 = \sin^{-1}\left(\frac{y_3 - y_2}{\sqrt{(x_3 - x_2)^2 + (y_3 - y_2)^2}}\right) - \sin^{-1}\left(\frac{y_3}{\sqrt{x_3^2 + y_3^2}}\right) \tag{3.9a}$$

$$\varphi_7 = \sin^{-1}\left[\frac{n_3}{n_4}\sin(\varphi_6)\right] \tag{3.9b}$$

$$P_4: \quad x_4 = x_3 + P_3P_4\cos\varphi_7, \; y_4 = y_3 + P_3P_4\sin\varphi_7 \tag{3.10}$$

Figure 3.3b is the projection of the light path on the section plane. It is easy to find the equivalent for each side of P_iP_j, which is like an equilateral non-closed polygon. The projected light will transmit with the same reflected angle, so it could be calculated by an iterative method.

If we know the points $P_{m\ n-1}$ $(x_{m\ n-1}, y_{m\ n-1})$ (where, the subscript 'm' represents the layer number (0 being in the fiber) and the subscript 'n' represents how many times the light is reflected in the fiber interior. Where P_4 in Fig. 3.3c is a contraction of "$P_{4\ 0}$".), point $P_{m\ n-2}(x_{m\ n-2}, y_{m\ n-2})$ and the angle φ_{n-1} between $P_{m\ n-1}P_{m\ n-2}$ and the x-axis, we can find the next reflecting point $P_{m\ n}$ $(x_{m\ n}, y_{m\ n})$ and the angle φ_n between $P_{m\ n}$ $P_{m\ n-1}$ and the x-axis. For the angles and coordinates in the fiber ($m = 0$), It is easy to count the reflecting points by rotating the coordinate using the following equation:

$$\varphi = \frac{y_{0(n-1)}}{\sqrt{y_{0(n-1)}^2 + x_{0(n-1)}^2}} \tag{3.11}$$

The new transmission triangle is $Q_{0\ n-1}$ $Q_{0\ n-2}$ $Q_{0\ n}$ instead of $P_{0\ n-1}$ $P_{0\ n-2}$ $P_{0\ n}$. The coordinates of points $Q_{0\ n}$ are as follows:

$$Q_{0n-2}: \quad Q_{0(n-2)}: u_{0(n-2)} = x_{0(n-2)}\cos\phi - y_{0(n-2)}\sin\phi, v_{0(n-2)} = y_{0(n-2)}\cos\phi + x_{0(n-2)}\sin\phi \tag{3.12}$$

$$Q_{0n}: \quad Q_{0n} = u_{0n} = u_{0(n-2)}, \; v_{0n} = -v_{0(n-2)} \tag{3.13}$$

Thus, we obtain the point $P_{0\,n}$ ($x_{0\,n}$, $y_{0\,n}$) by reversely rotating the coordinates:

$$P_{0n}: \quad x_{0n} = u_{0n}\cos(-\phi) - v_{0n}\sin(-\phi),$$
$$y_{0n} = u_{0n}\cos(-\phi) + v_{0n}\sin(-\phi) \tag{3.14}$$

These equations can be united and simplified as:

$$P_{0n}: \quad x_{0n} = -x_{0(n-2)}\cos(2\phi) + y_{0n}\sin(2\phi),$$
$$y_{0n} = y_{0(n-2)}\cos(-2\phi) + x_{0n}\sin(2\phi) \tag{3.15}$$

Then, using iterative method, all coordinates $P_{m\,n}$ (in layer m with n reflections) could be obtained by the Eqs. 3.6 ~ 3.10 and 3.15.

At the same time, in terms of the Eqs. 3.2a ~ 3.2d, and 3.15 above, the optical paths in Fig. 3.2a can be obtained by the following equations:

$$\Delta_1 = \tan\theta_2\sqrt{(x_1 - x_0)^2 + (y_1 - y_0)^2} \tag{3.16a}$$

$$\Delta_2 = \tan\theta_3\sqrt{(x_2 - x_1)^2 + (y_2 - y_1)^2} \tag{3.16b}$$

$$\Delta_3 = \tan\theta_5\sqrt{(x_3 - x_2)^2 + (y_3 - y_2)^2} \tag{3.16c}$$

$$\Delta_4 = \tan\theta_7\sqrt{(x_4 - x_3)^2 + (y_4 - y_3)^2} \tag{3.16d}$$

Later in this section, these equations will be used to simulate the light transmission.

3.1.2 *Attenuation and Absorption (Energy)*

There are three important factors influencing how the solar energy is separated, absorbed, and attenuated in the fiber: the Fresnel equations, the light absorbed by P3HT, and the light attenuated through each layer. All layers of the photovoltaic cell absorb some light. Although only the P3HT:PCBM layer converts that light into current, what is lost in the other layers cannot be used by the P3HT:PCBM layer. These factors must be included in the model in order to accurately determine photocurrent created by the device.

First of all, in terms of the Fresnel equations, the fractions of the incident power can be represented by the reflection coefficient R and the transmission coefficient T (refraction), where R and T depend on the polarization of the incident ray (*s*-polarized and *p*-polarized). The coefficients are given by:

$$T = 1 - R \tag{3.17}$$

$$R = \frac{1}{2}(R_S + R_P)$$
$$= \frac{1}{2}\left[\left(\frac{Cos\beta_i - \sqrt{n^2 - \sin^2\beta_i}}{\cos\beta_i + \sqrt{n^2 - \sin^2\beta_i}}\right)^2 + \left(\frac{-n^2\cos\beta_i + \sqrt{n^2 - \sin^2\beta_i}}{n^2\cos\beta_i + \sqrt{n^2 - \sin^2\beta_i}}\right)^2\right] \quad (3.18)$$

Where $\beta_{\mathrm{i}} = \sin^{-1}\frac{\mathrm{d_i}}{\Delta_{\mathrm{i}}}$ is the incident angle and $\mathrm{n} = \frac{\mathrm{n_t}}{\mathrm{n_i}}$ by Snell′s law. (3.19)

In terms of the formula for the distance from point to the plane:

$$d_i = \frac{\left|y_{i-1} + \frac{x_i x_{i-1}}{y_i} - \frac{x_i^2 - y_i^2}{y_i}\right|}{\sqrt{1 + \left(\frac{x_i}{y_i}\right)^2}}, i = 1, 2, 3, 4 \quad (3.20)$$

Secondly, we temporarily assume that there is a linear relation between absorption and attenuation with optical path. In other words, the quantity of light absorbed or attenuated is given by:

$$E_{absorbed} = a_i e^{\Delta_i} \quad (3.21a)$$

$$E_{attenuated} = b_i e^{\Delta_i} \quad (3.21b)$$

Where, i =1, 2, 3, 4 corresponds to the layers of the device: Fiber core, ITO, PEDOT, P3HT:PCBM, Δ_i is the optical path obtained by Eq. 3.16, and a_i,b_i are constants related to each layer's features.

As the incident angle is increased the power reaching the front face of the fiber decreases as in Fig. 3.4. The relationship of power entering fiber with incident angle is given by:

$$E_\theta = E_0 \cos\theta \quad (3.22)$$

Where $E_0 = \mu\pi r^2$ is the power entering at normal incidence,$\mu$is the power per unit area, and r is the radius of fiber.

3.2 Simulation Method

We utilized the iterative formulae by integrating the equations above:

$$E_n = E_{n-1}F_{mn}(1 - \sum_i C_{absorbed-i}(\Delta)) \quad (3.23)$$

$$E_{n-P3HT:PCBM} = E_{n-1}F_{mn}C_{abs-P3HT:PCBM}(\Delta) \quad (3.24)$$

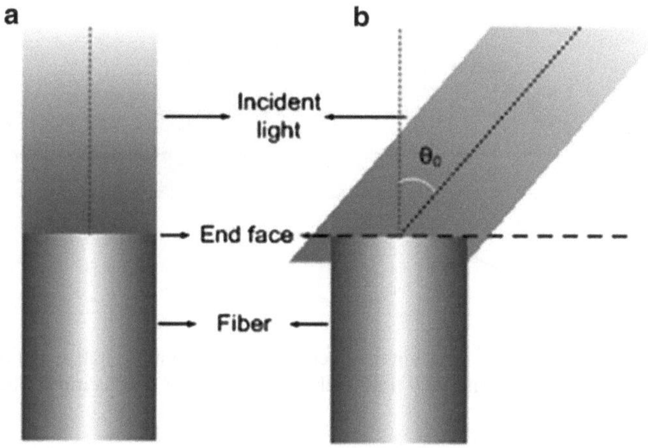

Fig. 3.4 (**a**) Illumination of the front of the fiber at normal incidence. (**b**) Illumination of the front of the fiber at incident angle , less power is incident on the fiber than with normal incident

Where

$$F_{mn} = \begin{cases} T_{nm} & transmission \\ R_{nm} & reflection \end{cases} \text{(Fresnel equations; nm represent from layer n to layer m)} \tag{3.25}$$

$$\sum_i C_{absorbed-i}(\Delta) = C_{abs-P3HT:PCBM}(\Delta) + C_{abs-ITO}(\Delta) + C_{abs-PEDOT}(\Delta) \tag{3.26}$$

$$\text{Initial condition :} \qquad \mathrm{E_1 = T_{01} E_0} \cos\theta, \;\; \mathrm{E_0} = \mu\pi \mathrm{r}^2 \tag{3.27}$$

A complete derivation is provided in the supplemental information available on line.

E_n is remaining energy at a given refracted point, which comes from the previous refracted point E_{n-1}. In this process, $E_{n-P3HT:PCBM}$ is the energy absorbed by P3HT: PCBM, which could be obtained by $C_{abs-P3HT:PCBM}(\Delta)$ (Energy absorbed is a function of optical path Δ in P3HT:PCBM) and F_{mn} (Fresnel equations). After entering the front face of the fiber, the light goes through the layers and reflects and refracts at each interface. Figure 3.2b shows that the light is separated at the interfaces with some photons either reflected or refracted resulting in two light rays within the device. Some of the power (in the form of absorbed photons) is lost at each interface and within each layer. In the P3HT:PCBM layer, those absorbed photons result in exciton generation. E_1 is the total power of the light entering the fiber relative to incident angle θ; E_0 is the power entering at normal incidence; $\mu$is the power per unit area, and r is the radius of fiber. $E_{absorbed}$ is light absorbed in

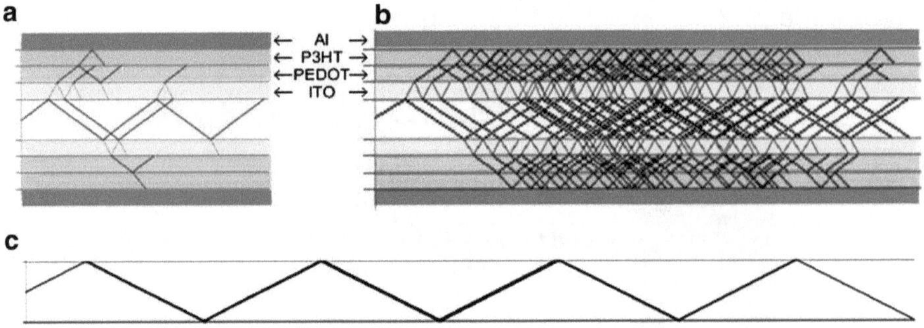

Fig. 3.5 Optical path of light ray entering the fiber. (**a**) Is the beginning status, and (**b**) is the latter case. (**c**) The light transmission in a real fiber cell (the diameter of fiber is much bigger than each layer of the cell)

each layer; T_{01} is the transmission coefficient of Fresnel equation, from air to fiber in Eq. 3.25. Further, the energy absorbed by the P3HT:PCBM layer in the fiber photovoltaic, can be obtained from the following equation:

$$Ratio_{absorbed} = \frac{E_{absorbed}}{E_{total}} \tag{3.28}$$

$$E_{absorbed} = \sum_{All\ \text{points in P3HT:PCBM}} E_{n-P3HT:PCBM} \tag{3.29}$$

The incident light is not a single light ray, but a light beam. For Fig. 3.2a, the incident point (x_0, y_0) is over a range of all rays incident on the front face of the fiber. The range of values is given by:

$$-\sqrt{r^2 - x_0^2} \le y_0 \le \sqrt{r^2 - x_0^2} \ \ where, -r \le x_0 \le r \tag{3.30}$$

The front face of the fiber (where the incident light enters) is divided into many discrete points (x_0, y_0) within the value range in Eq. 3.30. The total absorption result can be obtained in the form of an average from these discrete points Fig. 3.5.

In an experiment, there are usually extra conditions for the fiber cell that enhances the light absorption, such as the rear face of the fiber being mirrored by depositing a thin layer of aluminum on it. In Fig. 3.6b, the light will reflect on the rear face and make its way back along the fiber, giving it more time to be absorbed. In addition, other extra parameters like diameter and length of fiber must be considered in the simulation.

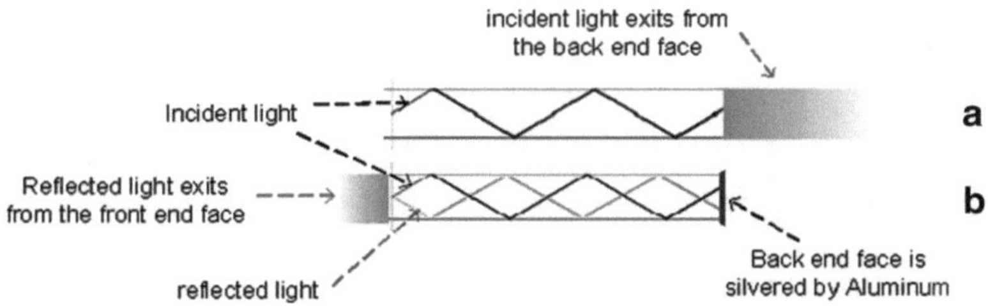

Fig. 3.6 (**a**) The fiber's rear face is not mirrored by aluminum, so the light exits from it. (**b**) The fiber's rear face is mirrored by aluminum, so the light reflects and travels back along the fiber, where it finally exits from the front face, this effectively doubles the length of the fiber in comparison to (**a**)

3.3 Dependence of Incident Angle and Architecture

Figure 3.2a shows a schematic of the device on which the calculations were performed. The calculated diameter was 1.4 mm. In Fig. 3.7a, Eq. 3.29 has been used to calculate the normalized power absorbed by the P3HT:PCBM layer as a function of the fiber length. Notice that maximum absorption at any angle occurs for the longest fibers, in this case 50 mm, as expected. However, as the fiber length decreases, the angle at which the maximum absorption occurs moves to larger values. For the shorter fiber lengths, it can also be seen that there are two peaks in the absorption. This trend suggests that there are two factors that influence the absorption and loss in the fiber photovoltaic cell which are based in geometrical optics. The first can be inferred from the long fibers. At normal illumination of the fiber face, some light is efficiently guided out the back of the fiber (or if the back of the fiber is mirrored, then it is reflected and exits out the front face). As the incident angle is increased (in the 50 mm long case maximum absorption is about 18°), absorption in the active layer is increased since more and more light is lost. The path length in the active layer is becoming long, and absorption is efficient. For very long fibers, this angle is related to the numerical aperture of the fiber. As the angle is increased further, to oblique illumination, the reflection plays a significant role in reducing the photons that couple into the front face of the fiber. Further, the flux is rolling off as the cosine of the angle of incidence.

In the case of shorter fibers, the situation at large incident angle is the same. However the angle at which the maximum loss occurs is shifted because the shorter fibers can transmit more of the light out the back (or front as mentioned above for mirrored backs). This means that more of the light must enter at angles in which more of the photons are lost "early" or near the front face of the fiber. For shorter fibers, this results in a fundamentally shorter optical path length in the absorber and consequently, less light is collected overall. Curiously, at intermediate angles, such as between 30° and 70° in the case of the 5 mm long fibers, there is a second feature

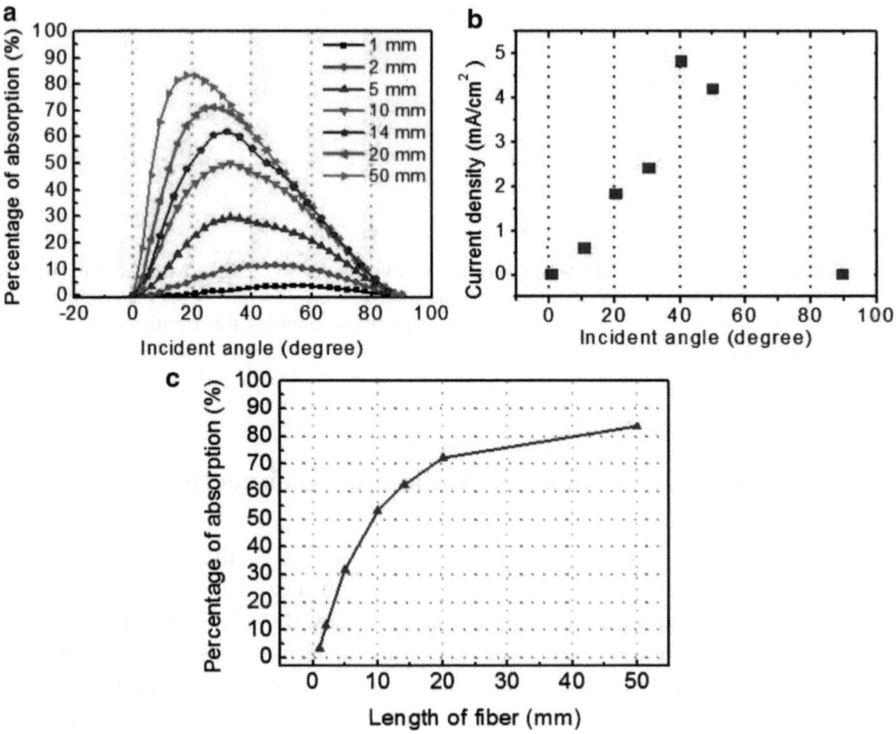

Fig. 3.7 (**a**) (Simulation) The light absorbed versus incident angle, by different fiber length (1 mm ~ 50 mm), with a diameter of 1.5 mm. (**b**) (Experiment) Short circuit current versus incident angle, with a length between 1 and 2 mm, and a diameter of 1.4 mm. (**c**) (Simulation) In terms of (**a**), the percentage of absorption in fiber with the optimum incident angles for different fiber lengths, with a diameter of 1.4 mm, incident angle $\theta_0 = \pi/4$, $\varphi_1 = \pi/2$

that indicates enhanced absorption. This is likely to be related to a second confined mode of the fiber, probably a radial mode. This mode is in fact expected to be present in all lengths of the fibers, but is dominated by the large "linear" absorption response in the case of the long fibers and only seen in the shorter fibers. It is important to keep in mind that these simulations are performed for 1.4 mm diameter fibers, and will look different for different diameters as we examine below.

In Fig. 3.7b we compare our experimental findings of current generation in a 1.4 mm diameter fiber that is between 1 mm and 2 mm, to the simulations of Fig. 3.7a. The fiber photovoltaics used in this study were fabricated by using the method outlined by J. Liu et al. [4, 5] and were tested by an AM1.5g standard (Oriel) operating with an illumination intensity of 100 mW/cm^2. The thicknesses of each layer on the fiber are: PEDOT:PSS(~150 nm), P3HT:PCBM (~300 nm), LiF (0.3–0.4 nm), and Al(~100 nm). The maximum absorption peak corresponds to incident angle of 55° in agreement with the simulation of Fig. 3.7a. From Fig. 3.7c, it can be seen that the longer fiber length is beneficial to the absorption of light,

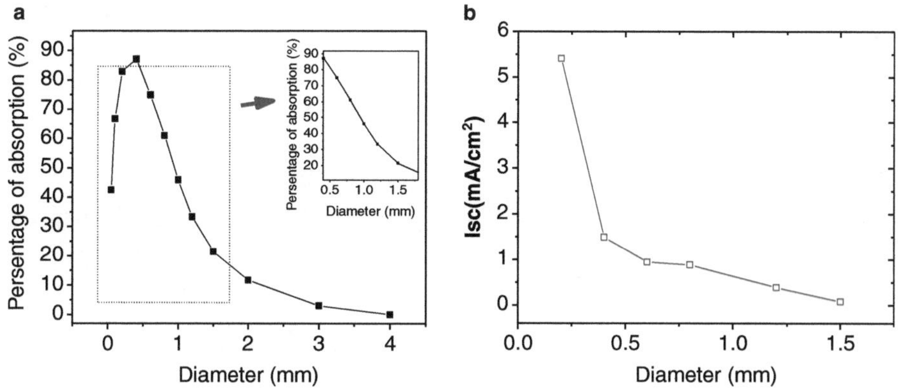

Fig. 3.8 (**a**) (Simulation) Light absorbed versus fiber diameter, with a length of 14 mm (**b**) (Experiment) Efficiency versus fiber diameter, with a length of 14 mm

but its effect will approach a constant (maximum absorption). In other words, it is useless to extend fiber length too far, since the fiber has absorbed all light power beyond a certain length.

We can also examine the effect of the diameter on the light absorption and loss in the fiber geometry. Figure 3.8b shows the experimentally determined current densities as the fiber diameter is varied (for normally incident light). The current generation increases significantly as the diameter of the fiber is decreased, which corresponds to the absorption predicted by the model, shown in Fig. 3.8a. This is reasonable since the number of ray reflections in the fiber is inversely proportional to the fiber diameter as is expected from geometrical optics. Thus, if the diameter is too big, more light would exit from the rear face for a given finite length.

However, since the fiber' tip surrounded by ITO is used as anode connecting circuit in Fig. 3.9a, i.e. doesn't include aluminum on outer surface, incident light would pass through the fiber glass without absorption. This leads to a shift of optimum incident angle as shown in Fig. 3.9a, namely longer exposed tip smaller optimum incident angle.

The above calculations were performed by summing the light incident across the entire face of the fiber. This corresponds to the spread of angles and incident points present in the diffuse standard used in our experiments. It is instructive to look at the absorption at each point across the fiber face to understand how each ray is contributing to the absorption. As shown in Fig. 3.10, for light incident of a specific angle in θ_0 and ϕ_1 on the middle of the front face of the fiber, the absorption effect is better than that on the outer edges. High absorption is shown in green, and lower absorption is shown in other colors. The azimuthal angle ϕ_1 is set to $\pi/2$ for the example shown, which causes the transmittance (T) to become small along the "y" edge.

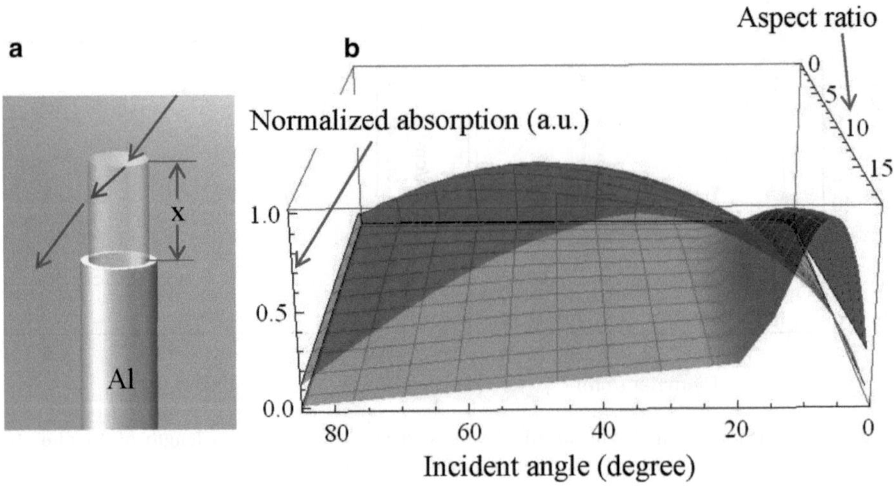

Fig. 3.9 (**a**) Light lost at the exposed tip with length x (**b**) Relationship between the PTL length, incident light angle and light absorption. *Blue curve* is absorption of x = 0, and *green curve* is that of x = 3 times of diameter

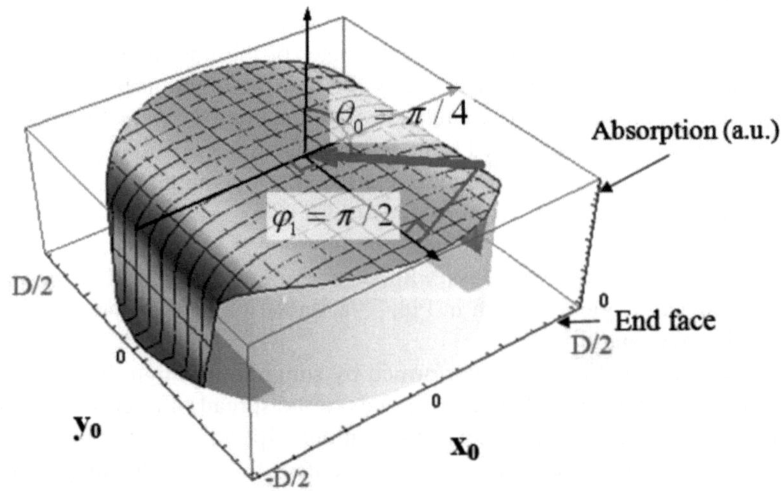

Fig. 3.10 (Simulation) Absorption for the incident point (x_0, y_0) with incident angles $\theta_0 = \pi/4$, $\phi_1 = \pi/2$. (Refer to Fig. 3.2) The under surface x0-y0 plane is the end face of fiber. The *green* part is the area of best absorption, the *red* is the worst. Where, D is diameter of end face

References

1. Y. Li, W. Zhou, D. Xue, J. Liu, E. Peterson, W. Nie, D. Carroll, Origins of performance in fiber-based organic photovoltaics. Appl. Phys. Lett., **95**, 203503 (2009)
2. Y. Li, W. Zhou, D. Xue, J.W. Liu, E.D. Peterson, W.Y. Nie, D.L. Carroll, Origins of performance in fiber-based organic photovoltaics. Appl. Phys. Lett. **95**, 203503 (2009)
3. S. Curran, J. Talla, S. Dias, J. Dewald, Microconcentrator photovoltaic cell (the m-C cell): Modeling the optimum method of capturing light in an organic fiber based photovoltaic cell. J. Appl. Phys. **104**, 064305 (2008)
4. J.W. Liu, M.A.G. Namboothiry, D.L. Carroll, Fiber-based architectures for organic photovoltaics. Appl. Phys. Lett. **90**, 063501 (2007)
5. J.W. Liu, M.A.G. Namboothiry, D.L. Carroll, Optical geometries for fiber-based organic photovoltaics. Appl. Phys. Lett. **90**, 133515 (2007)

Chapter 4
Spectral Response

4.1 Transfer Matrix Model for Optical Confinement Geometry

The fiber-based solar cell is an example of a three dimensional photovoltaic architecture that uses the natural mode structure of optical fibers to enhance light capture (Reprinted from [1], Copyright (2011), with permission from Elsevier). In this work we explore the spectral response of such cells when the thickness of the absorbing layer is varied. We demonstrate two important consequences associated with this architecture. The first is that fiber-based devices generally require a thinner active layer than the analogous planar structure. This helps to avoid exciton recombination and reflection loss in the geometry. Secondly, the geometry exhibits a broader absorption and external quantum efficiency than its planar counterpart. We interpret this as being due to enhanced absorption of charge transfer excitons.

As discussed in Sect. 1.3.1, comparing with some approaches including optical spacer, tandem structure and inverted architecture, three dimensional light capturing architectures are another class of solutions that can address materials shortcomings [2–5]. An example of such 3D architectures, the end illuminated fiber geometry [6, 7] was used to essentially trap incoming light into a fiber-optic coated with a photoconversion material. As the light proceeds down the fiber, it is absorbed by the photoconversion material resulting in extraordinarily high absorption by the device. For fiber devices, and indeed for much other 3D architecture, one might guess that material costs increase rapidly in this case because the surface area of the fiber is greater than the area of the fiber face that it replaces. Moreover, it might be expected that the performance enhancement in such architecture is limited by the same absorption window as presented by the analogous planar device. In this section, we use a combination of experiment and calculation to examine this expectation in detail. The surprising outcome is that the ideal absorber thickness for the fiber device is substantially thinner than for the analogous planar device. Furthermore, the window of absorption and power generation is broadened to a surprising degree.

In modeling the behavior of fiber-based photovoltaics, both the optical line and electrical wave must be taken into account in terms of the media's length scales.

Y. Li, *Three Dimensional Solar Cells Based on Optical Confinement Geometries*, Springer Theses, DOI 10.1007/978-1-4614-5699-5_4,

That is, when the light propagates in a fiber with diameter of ~1 mm, the optical path is much larger than that of the wavelength of visible light, so the optical path can be calculated by a ray tracing model as we have reported before [7]. This model can simulate the absorption of the device with varying incident angle, incident position, fiber length and ratio, etc. When the light transmits into the stacked nanometer-scale layers, light's interference influences the optical field distribution and energy absorption in multi-layers. In this work, we use the fiber-based solar cell as an example of the 3D photovoltaic architecture to explore the spectral response in such devices as the absorber thickness is varied.

In terms of the Eq. 1.53 in Chap. 1, the TE wave and TM wave of incident light are equal to half of solar energy, such that we can calculate their absorption separately and sum them together.

$$Q(x) = Q_{TE}(x, \lambda, I_{TE}(\lambda)) + Q_{TM}(x, \lambda, I_{TM}(\lambda)) \tag{4.1}$$

By means of Eq. 4.1, the energy absorption in each layer could be worked out, and their ratio A_i (layer i = 1, 2, ...m) to is independent with incident light intensity, shown in Eq. 4.2. Thus, for fast calculation, this ratio is helpful to find the energy absorption in certain layer via iterative multi-reflected propagation in fiber chamber.

$$A_i = Q_i \Big/ \sum_{i=1}^{m} Q_i \tag{4.2}$$

Figure 4.1c represents the process that light enters into fiber and reflects many times. The materials including air, fiber, transparent anode, hole transport layer, active layer, electron transport layer, and cathode, which are numbered as 0, 1, 2, ...*i*, ..., and *m successively*, where *a* represents the active layer. The ray tracing method [7, 8] is used to calculate the position and the angle of each reflection. However, since the fiber cannot bend over short scale, it is reasonable to assume the incident angles at each reflection are same, i.e. θ_i = constant. At the first reflection as shown in Fig. 4.1c, the light absorption from active layer in stacked films could be written as Eq. 4.3.

$$Q_{a\lambda}^{[1]}(\lambda) = \alpha_{a\lambda} n_{a\lambda} A_{a\lambda} T_{m\lambda} T_{0\lambda} I_{0\lambda} e^{-i\alpha_{1\lambda} d/\cos(\vartheta_1)} \tag{4.3}$$

Where d, λ; are fiber diameter, wavelength respectively. $I_{O\lambda} = (I_{TE} + I_{TM})/2$ is incident light intensity combined by TE wave and TM wave which is obtained by Eq. 1.53 in Chap. 1.

The corresponding transmittance between air and fiber is $T_{O\lambda} = |t_{Ol}|^2$, shown in Eqs. 4.4 ~ 4.5; The refractive index $n_{a\lambda}$ is from the complex refractive index $\tilde{n}_{i\lambda} = n_{i\lambda} + ik_{i\lambda}$; T_m, R_m and $A_{a\lambda}$ are the transmittance and reflectance between fiber and stacked multi-layers that decided by the Fresnel equation, and the ratio of the energy absorbed by active layer and that by all stacked multi-layers, which are

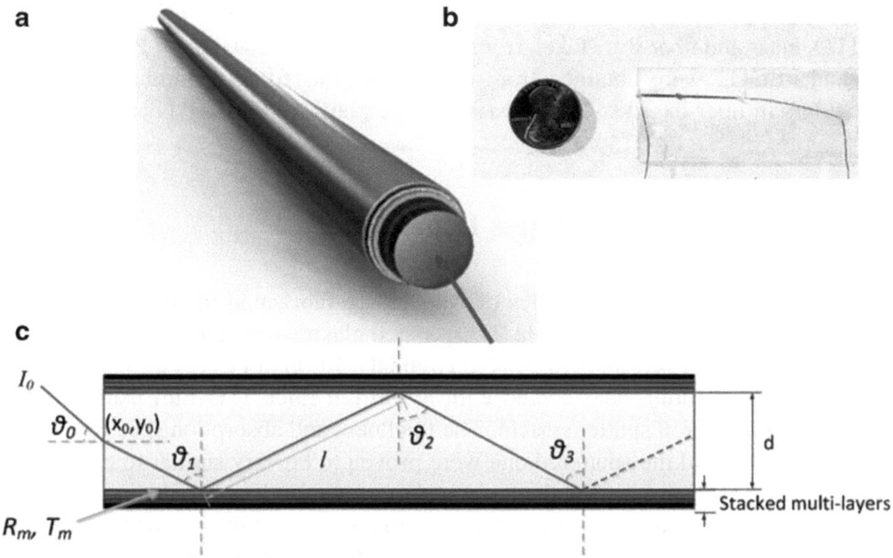

Fig. 4.1 (**a**) Schematic diagram of light entering into fiber-based solar cell. (**b**) Image of the fiber-based solar cell device. (**c**) Light incidence and propagation in fiber. At position (x_O, y_O), sunlight enters into fiber (diameter *d*) from the left side with incident angle θ_O. Light reflects on inner surface of fiber with incident angle θ_1. The reflectance and transmittance factors between fiber and stacked multi-layer are R_m, T_m in Eqs. 1.17 and 1.18. $l = d/sin(\theta_1)$ is the optical path between two reflections

derived from a transfer matrix method [9–12]. $\alpha_{a\lambda} = 4\pi k_{a\lambda}/\lambda$ is the absorption coefficient in active layer.

$$\text{TE Wave } r_{jk} = \frac{q_j - q_k}{q_j + q_k}, t_{jk} = \frac{2q_j}{q_j + q_k} \tag{4.4}$$

$$\text{TM Wave } r_{jk} = \frac{\tilde{n}_k^2 q_j - \tilde{n}_j^2 q_k}{\tilde{n}_k^2 q_j + \tilde{n}_j^2 q_k}, t_{jk} = \frac{2\tilde{n}_k \tilde{n}_j q_j}{\tilde{n}_k^2 q_j + \tilde{n}_j^2 q_k} \tag{4.5}$$

Where
$$q_j = \tilde{n}_j \cos\theta_j = \tilde{n}_j(1 - (n_0 \sin\theta_0/n_j)^2)^{1/2} \tag{4.6}$$

The subscript *j*, *k* are the numbers of incident and refractive layers. In the straight fiber, the total absorption with n reflections can be written as:

$$\begin{aligned} Q_{a\lambda}^{total} &= \sum_n Q_{a\lambda}^{[n]} = Q_{a\lambda}^{[1]} + (1 - T_m)Q_{a\lambda}^{[1]}e^{-i\alpha_1 l} + \cdots\cdots \\ &+ (1 - T_m)^{n-1}Q_{a\lambda}^{[1]}e^{-i(n-1)\alpha_1 l} \\ &\xrightarrow{\text{if } n -> \infty} \frac{Q_{a\lambda}^{[1]}}{1 - (1 - T_m)e^{-i\alpha_1 l}} = \frac{Q_{a\lambda}^{[1]}}{1 - R_m e^{-i\alpha_1 l}} \end{aligned} \tag{4.7}$$

Several parameters are necessary in this model. The complex refractive indexes of ITO, glass and fiber were taken from reference [13] and the data of P3HT:PCBM, Al and PEDOT were obtained from references [14–16]. We simulated the light behaviors in fiber-based solar cell based on the platform OPVAP [12].

4.2 Active Layer Thickness Dependence

As shown in Table 4.1, two series of devices are fabricated to compare with the simulation above. Indium tin oxide (ITO) coated glasses were purchased from Delta for the planar devices. Fibers (0.39 NA, High OH, 1,000 μm Core) were stripped of their polymer cladding and coated with a 100 nm thick ITO film using a radio frequency magnetron sputter system. The thickness and absorption spectrum of the purchased ITO and the sputtered one were proven to be very similar to each other. The fibers and substrates were then exposed to ozone for 30 min (rotating the fibers three times per 30 min). Subsequently, by a dip coating (for fiber by an Opticer H10 Dip Coater), and a spin coating (for planar device) procedures, Poly(3,4-ethylene-dioxythiophene)-poly(styrenesulfonate) (PEDOT:PSS, Clevios P) and Poly 3-hexyl thiophene:Phenyl-C_{61}-butyric acid methyl ester (P3HT:PCBM = 1:0.8 in WT) were deposited on the substrates. Thicknesses of P3HT:PCBM were controlled by varying the concentration of solution when other parameters were kept constant (e.g. spinning speed for planar device and withdraw speed for fiber-device). Finally, Al electrodes were deposited via thermal evaporation at the pressure of 10^{-6} torr. (Fiber rotating was controlled by an Opticer C10 rotator in vacuum.) The architecture and corresponding thickness of all devices are shown in Table 4.1. The thicknesses were from scanning electron microscope (SEM) images. Devices were annealed at 150°C for 7 min. Then, the devices were tested in air using an Oriel AM1.5 g standard Newport # 96000 Solar Simulator. Current–voltage characteristics were measured using a Keithley 236 source-measurement unit. The external quantum efficiency (EQE) was measured on a Newport Merlin lock-in amplifier by shinning a Newport 300 W Xenon light source through a Newport Cornerstone 260 Monochromator.

As mentioned before, fiber-based organic photovoltaics can absorb a significant amount of the incident light without reflection loss and also avoids high recombination rate using a thin active layer [17, 18]. In Fig. 4.2, the calculated Jsc (without recommendation rate) versus thickness (two solid lines) are plotted directly from the combined modeling mentioned above. The measured values of Jsc for both device types are also plotted. Clearly, the absorption of the fiber-based cell is higher

Table 4.1 Structure and thickness of devices studying spectral response in fiber cell

	Glass	ITO	PEDOT	P3HT:PCBM	Al
Planar cell	1 mm thickness substrate	100 nm	40 nm	20 ~ 280 nm	100 nm
Fiber cell	1 mm diameter fiber	100 nm	40 nm	20 ~ 280 nm	100 nm

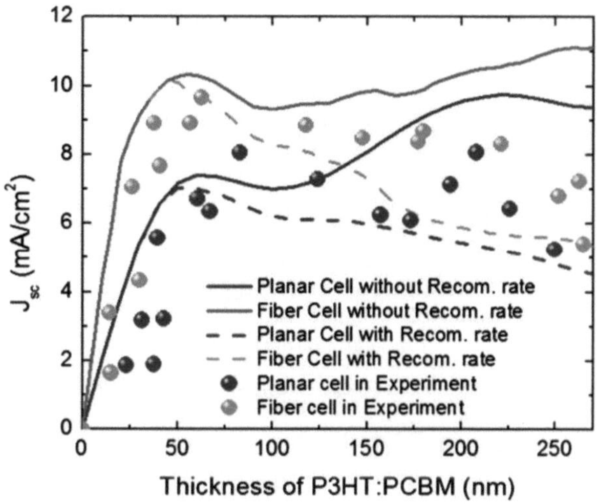

Fig. 4.2 (**a**) Thickness dependence of short-circuit current density J_{sc} for planar cell and fiber cell. *Sphere points* is experiment data; *Solid film* is simulation result without recombination rate; *Dashed line* is simulation results with Langevin recombination efficiency 100% [19]

than that of the planar cell for any thickness of active layer. Moreover, the first peak (~50 nm in absorber thickness) of the fiber-based cell's absorption has nearly reached saturation, significantly higher than that of the planar cell. This suggests that such geometrical approaches can be quite efficient using very thin active layers to reach a full absorption.

To further account for the recombination rate, we used a Langevin recombination efficiency factor of 1 (the range is from 0 to 1, and 1 corresponds the maximum recombination rate) [11, 19] to obtain the Jsc with maximal recombination. (two dashed curves). We note that the experimental data for both the planar and fiber-based cells yield a Jsc that is below that of the predicted values without recombination, but larger than expected if a maximal recombination is considered. This suggests that both cells are subject to some recombination loss. However, clearly the fiber-based devices are capable of producing more current with a thinner absorber.

4.3 Absorption and EQE

The efficiency of planar cells and fiber-based cells are generally at 70 and 100 nm, respectively. Figure 4.3 illustrates the energy absorption distribution and corresponding EQE for the devices. Figure 4.3a shows that reflection leads to large energy losses (yellow area) in the planar solar cell. However, as shown in Fig. 4.3b, fiber-based solar cells can absorb this reflection by continuous propagation down the fiber. Using the AM1.5 g solar spectrum [20], EQE were obtained and are shown in Fig. 4.3c, d. Generally speaking, experimental data fit the simulation well, but in the fiber architecture, a broadening of the absorption, and thus in the power generation, is observed in the experimental data. Note that the absorption

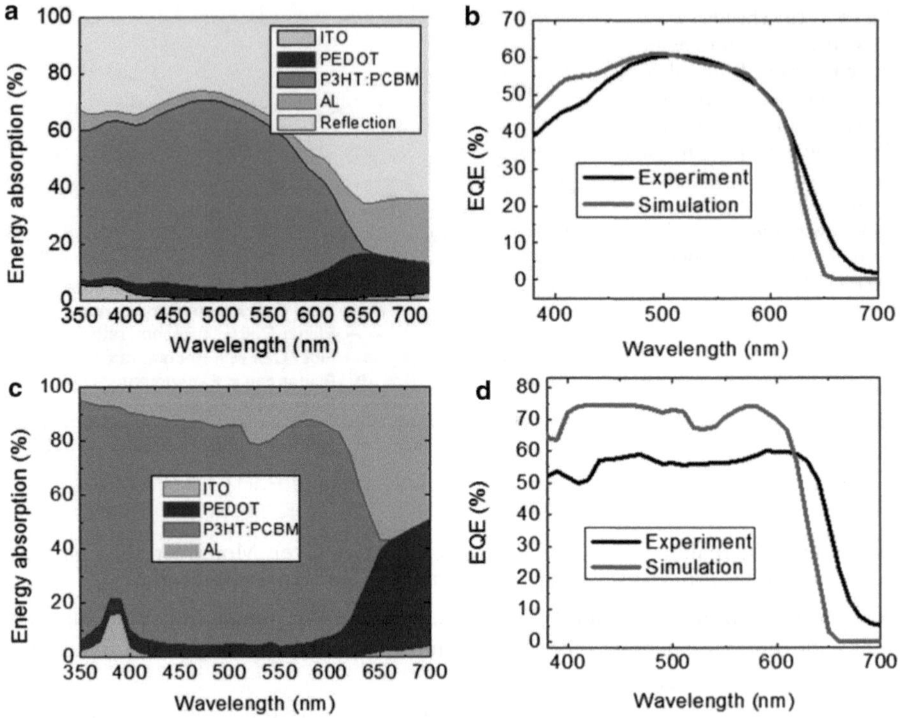

Fig. 4.3 (**a, b**) Redistribution of incoming irradiation on reflection and layerwise absorption for planar cell and fiber cell, respectively. (**c, d**) The comparison of EQE spectrum of experiment and simulation for planar cell and fiber cell, respectively. The selected thicknesses of P3HT:PCBM for planar cell and fiber cell are 100 and 70 nm, respectively. The incident angles for planar cell and fiber cell are $\theta_0 = 0°$, and $\theta_0 = 10°$, respectively

edge, and turn-on power in the EQE for the planar device occurs at ~650 nm, whereas for the fiber-based device, power is being generated as low as 675 nm and lower. This peculiar broadening of the spectral response is seen in all fiber-based devices tested. While the origin of this broadening is still under investigation, a possible explanation is the contribution of charge transfer exciton absorption (CTEA) [21, 22]. Through this mechanism, the absorption tail could be extended to nearly 900 nm (1.38 eV) [23]. This phenomenon increases through morphological structuring brought on by the annealing process [24, 25]. Since the CTEA is weaker than donor-acceptor exciton absorption (DEA), especially in the infrared wavelength, this part of the energy cannot be absorbed in only one reflection as in the planar solar cells. In contrast, in fiber-based solar cells, the continuous reflections enhance the influence of CTEA. In addition, since fiber-based cell fabrication is not perfect, the EQE of experiment has not reached the level of simulation but it does maintain a similar shape.

In summary, using the fiber-based solar cell architecture as an example of 3D organic photovoltaics, we have explored their spectral performance as the thickness of the absorber is varied. We have shown that maximal current densities and performance is obtained in such architectures at lower absorber thicknesses than in the planar configuration. We have also identified a phenomenological broadening of the absorption and power generation which we suggest is due to CTEA brought on by the extended optical paths in the fiber-based devices.

References

1. Y. Li, H. Huang, M. Wang, W. Nie, W. Huang, G. Fang, D.L. Carroll, Spectral response of fiber-based organic photovoltaics. Sol. Energ. Mat. Sol. C. **98**, (2011). doi:10.1016/j.solmat.2011.10.033
2. Y. Li, M. Wang, H. Huang, W. Nie, Q. Li, E.D. Peterson, R. Coffin, G. Fang, D.L. Carroll, Influence on open-circuit voltage by optical heterogeneity in three-dimensional organic photovoltaics. Phys. Rev. B. **84**, 085206 (2011)
3. Y. Li, E.D. Peterson, H. Huang, M. Wang, D. Xue, W. Nie, W. Zhou, D.L. Carroll, Tube-based geometries for organic photovoltaics. Appl. Phys. Lett. **96**, 243505 (2010)
4. H. Huang, Y. Li, M. Wang, W. Nie, W. Zhou, E.D. Peterson, J. Liu, G. Fang, D.L. Carroll, Photovoltaic-thermal solar energy collectors based on optical tubes. Sol. Energy **85**, 450–454 (2011)
5. Y. Li, W. Nie, J. Liu, A. Partridge, D.L. Carroll, The optics of organic photovoltaics: Fiber-based devices. IEEE J. Sel. Top. Quant. Electron. **16**, 1827–1837 (2010)
6. B. O'Connor, K.P. Pipe, M. Shtein, Fiber based organic photovoltaic devices. Appl. Phys. Lett. **92**, 193306 (2008)
7. Y. Li, W. Zhou, D. Xue, J. Liu, E.D. Peterson, W. Nie, D.L. Carroll, Origins of performance in fiber-based organic photovoltaics. Appl. Phys. Lett. **95**, 203503 (2009)
8. S. Curran, J. Talla, S. Dias, J. Dewald, Microconcentrator photovoltaic cell (the m-C cell): Modeling the optimum method of capturing light in an organic fiber based photovoltaic cell. J. Appl. Phys. **104**, 064305 (2008)
9. D.W. Sievers, V. Shrotriya, Y. Yang, Modeling optical effects and thickness dependent current in polymer bulk-heterojunction solar cells. J. Appl. Phys. **100**, 114509 (2006)
10. L.A.A. Pettersson, L.S. Roman, O. Inganas, Modeling photocurrent action spectra of photovoltaic devices based on organic thin films. J. Appl. Phys. **86**, 487–496 (1999)
11. R. Hausermann, E. Knapp, M. Moos, N.A. Reinke, T. Flatz, B. Ruhstaller, Coupled optoelectronic simulation of organic bulk-heterojunction solar cells: Parameter extraction and sensitivity analysis. J. Appl. Phys. **106**, 104507 (2009)
12. Open Photovoltaics Analysis Platform (OPVAP) by Yuan Li, USA. www.OPVAP.com
13. H. Hoppe, N.S. Sariciftci, D. Meissner, Optical constants of conjugated polymer/fullerene based bulk-heterojunction organic solar cells. Mol. Cryst. Liq. Cryst. **385**, 233–239 (2002)
14. F. Monestier, J.J. Simon, P. Torchio, L. Escoubas, F. Florya, S. Bailly, R. de Bettignies, S. Guillerez, C. Defranoux, Modeling the short-circuit current density of polymer solar cells based on P3HT: PCBM blend. Sol. Energ. Mat. Sol. C. **91**, 405–410 (2007)
15. Refractive Index. Info. database http://refractiveindex.info
16. M. Caglar, S. Ilican, Y. Caglar, F. Yakuphanoglu, The effects of Al doping on the optical constants of ZnO thin films prepared by spray pyrolysis method. J. Mater. Sci-Mater. El. **19**, 704–708 (2008)

17. A. Pivrikas, G. Juscaronka, A.J. Mozer, M. Scharber, K. Arlauskas, N.S. Sariciftci, H. Stubb, R.O. Sterbacka, Bimolecular recombination coefficient as a sensitive testing parameter for low-mobility solar-cell materials. Phys. Rev. Lett. **94**, 176806 (2005)
18. F. Yang, M. Shtein, S.R. Forrest, Controlled growth of a molecular bulk heterojunction photovoltaic cell. Nat. Mater. **4**, 37–41 (2005)
19. G. Juska, K. Arlauskas, J. Stuchlik, R. Österbacka, Non-Langevin bimolecular recombination in low-mobility materials. J. Non-Cryst. Solids **352**, 1167–1171 (2006)
20. ASTM Standard G173, 2003e1, *Standard Tables for Reference Solar Spectral Irradiances: Direct Normal and Hemispherical on 37° Tilted Surface* (ASTM International, West Conshohocken, PA, 2003). doi: 10.1520/G0173-03E01, www.astm.org
21. X.J. Wang, E. Perzon, J.L. Delgado, P. de la Cruz, F.L. Zhang, F. Langa, M. Andersson, O. Inganas, Infrared photocurrent spectral response from plastic solar cell with low-band-gap polyfluorene and fullerene derivative. Appl. Phys. Lett. **85**, 5081–5083 (2004)
22. X.Y. Zhu, Q. Yang, M. Muntwiler, Charge-transfer excitons at organic semiconductor surfaces and interfaces. Accounts. Chem. Res. **42**, 1779–1787 (2009)
23. Y. Chia-Ming, T. Pei-Yu, H. Sheng-Fu, L. Kuan-Chen, T. Shin-Rong, M. Hsin-Fei, S. Jow-Tsong, S. Ching-Fong, Infrared photocurrent response of charge-transfer exciton in polymer bulk heterojunction. Appl. Phys. Lett. **92**, 083504 (2008)
24. R.A. Marsh, J.M. Hodgkiss, S. Albert-Seifried, R.H. Friend, Effect of annealing on P3HT: PCBM charge transfer and nanoscale morphology probed by ultrafast spectroscopy. Nano Lett. **10**, 923–930 (2010)
25. I.W. Hwang, D. Moses, A.J. Heeger, Photoinduced carrier generation in P3HT/PCBM bulk heterojunction materials. J. Phys. Chem. C **112**, 4350–4354 (2008)

Chapter 5
Individual Optical Confinement Geometry Device

In terms of structure and feature, the optical confinement geometry solar cell is classified as individual OCGPV, integrated OCGPV, and hybrid OCGPV, in which the individual OCGPV is focused on devices based on single fibers or tubes. Since the last two types, integrated OCGPV and hybrid OCGPV, are developed and studied from the single device, individual OCGPV shows some common characteristics with the other two types, such as incident angle dependence, spectral response, and equivalent circuit. In this section, the individual OCGPV is introduced via two examples of a fiber based solar cell with dip coating ITO and ITO-free Tube-based solar cell before discussing more industrialized integrated OCGPV and more special features of hybrid OCGPV. In the research of individual OCGPV, an important issue is choosing transparent electrode surround fiber or tube, because it is very hard to make smooth films on this 3D architecture surface. From year 2007 to 2012, we studied several methods for depositing anode: sol–gel ITO by dip coating, conductive PEDOT as the replacement of ITO, and the traditional method by sputtering ITO. In addition, some tools and instruments are invented to serve this 3D fabrication process in which most of them are shown in Sect. 9.1.

5.1 Fiber Based Solar Cell with Dip Coating ITO

Conformal, high conductivity thin films of indium tin oxide (ITO) have been deposited utilizing a relatively low temperature, ambient, multi-layer dip coating process (Reprinted with permission from [1], Copyright (2009), American Institute of Physics). Using standard alcohol solutions of indium chloride and stannic chloride to form ultra-thin layers, consecutive multi-layer deposition of ITO onto telecommunications grade glass fibers (as a model system) showed excellent control over grain morphology in the film as determined by electron microscopy, and a linear relationship between thickness and overall fiber conductivity. Ray tracing, coupled with a transfer matrix formalism, was used to numerically simulate the

Y. Li, *Three Dimensional Solar Cells Based on Optical Confinement Geometries*, Springer Theses, DOI 10.1007/978-1-4614-5699-5_5,

effects of film thickness on the optical wave-guiding nature of the conducting layer. The simulations, carried out for a generic film morphology, show that a significant fraction of the optical energy, coupled into the fiber face, is transmitted into the film at the thicknesses studied. These results were then used to estimate an upper limit of optical power transmission provided for the generic system. From this, a comparison between the optical performance of sputter deposited and multi-layer conducting oxide films in a device configuration could be made. Organic photovoltaic devices, using both sputter deposited and multi-layer conductors on optical grade fibers, were fabricated and tested. Both compared favorably to the numerical simulations, suggesting that the overall, long range performance between the multi-layer and sputtered films are comparable as cathodes for such conformal devices.

Increasingly, applications of organic devices have turned to conformal (three-dimensional) shapes. Such applications include fiber-based photovoltaics, light emitting diodes, and optical amplifiers [2–6]. Typically, device architectures using organic layers require a transparent conductor. This cathode should be an evenly applied thin film with high conductivity and a specific grain structure that results in reduced light scattering. In many of these applications, indium tin oxide (ITO) has been widely used as the transparent electrode due to the high optical transparency in the visible spectral range and the high electric conductivity ($<100\ \Omega/m^2$). For flexible substrates, ITO can be applied in very thin, high quality films, onto a flexible backing like polymethylmethacrylate, or other transparent polymer [7, 8]. However, when the substrate shape is a fixed, complex, three dimensional shape, application of a uniform film becomes very difficult due to shadowing, the tendency of solutions to fill crevasses unevenly, and exaggerated grain growth in regions of high surface tension. But, device performance demands relatively even thicknesses and a uniform grain structure in the film, otherwise conduction "hot spots" can occur, along with localized light trapping.

Various methods, like sputtering [9] evaporation [10], spray pyrolysis [11] and screen printing [12], have been explored to prepare high quality ITO thin films. There are few reports of ITO film coating on substrates with different shapes. Most high performance device data comes from planar substrates where the ITO film morphology and its dependence on film thickness can be readily controlled. To circumvent the difficulties presented by the ITO film morphology on complex shapes, researchers such as Brendan O'Connor et al. have used vacuum deposited thin metal films as the transparent conductor for both OLED (organic light emitting diodes) [2] and OPV (organic photovoltaics) [13] based on fiber architectures. Compared to these non-ITO electrodes, ITO has a higher work function, which is important to many device applications such as OPVs and OLEDs since it tends to block electrons. Very recently, Benjamin Weintraub et al. prepared (dye sensitized solar cells) DSSCs on sputtered-ITO coated optical fibers [3, 6]. But, this requires relatively sophisticated substrate manipulation in vacuo during deposition, making it less attractive for large scale production of substrates. Using sol–gel methods, ITO thin films can also be deposited onto optical fibers, and organic solar cells have been successfully fabricated onto dip coated ITO fibers [4, 5]. In these cases, film thickness can be generally increased by applying multiple layers of the ITO film, thereby increasing its conductivity. However, quantitative correlations between morphology, thickness, resistivity, and light trapping of multiple dip-coated ITO films on such three dimensional surfaces are not clear. Thus, the technical challenges of uniform transparent cathode formation on conformal pieces remain a major stumbling block to realizing truly high performance, three-dimensional devices in this context.

In this study, we examine the preparation of ITO thin films deposited by a solution-based, multi-layer dip-coating process onto optical fibers (used as our model system). The approach uses a standard sol–gel preparation and the consecutive deposition of very thin layers to create the cathode film. Previous work in this area has focused on the application of a relatively thick set of layers, using high concentrations of the sol-gels, and resulting in films with a high degree of variability. However, when a low concentration sol–gel is applied in a slow and controlled multi-layer process, the formation of very fine grain structure during deposition is possible. In this work, overall film thickness and linear conductivity change with the number of layers characterized and compared to sputter deposited films. Micro-morphology was determined by electron microscopy. Numerical simulations were performed to estimate the optimal optical performance of the coating as a function of thickness for a generic film morphology, as applied to a fiber geometry. These simulations were then used to compare the photovoltaic performance of multi-layer and sputter deposited films in a standard polymer solar cell fabricated by putting the thin bulk-heterojunction film composed of regioregular poly(3-hexylthiophene) (P3HT) and chemically modified fullerene derivative [6,6]-phenyl-C61-butyric acid methyl ester (PCBM) as the solar absorbing layer on the conductive fibers and then covering with an aluminum contact by thermal evaporation. It was determined that the layer-by-layer coated ITO film can be completely comparable to the ITO films from sputtering coating in device performance. While this result isn't surprising for planar devices, the technique's ability to yield high performance conformal coatings, with long range conductivity and transparency, was unexpected and suggests that the interfaces between thin layers do not significantly influence overall film performance. Thus, the approach appears to be an excellent general approach to conformal cathode formation for such devices.

5.1.1 Fabrication of Fiber-Based Solar Cell

5.1.1.1 Dip-Coating Process

In this work, telecommunications quality optical fibers were used for a model three dimensional system. While it is recognized that optical fibers do not represent the full range of complexity that we might seek to coat using these techniques, given the growing importance of fiber-based devices, they do present an important starting point for these studies. To prepare the fiber substrates, the jacket of multimode fibers (BFH37, high OH, 0.2 mm diameter from Thorlabs) was stripped off and the inner, hard polymer cladding was removed. Then the fiber core was cleaned ultrasonically with acetone, and dried at 100°C for 15 min. Fibers were examined microscopically to ensure that the outer cladding was removed completely and the starting surface was free of defects. Fiber lengths of several centimeters were used for all comparisons.

ITO sol–gel solutions were prepared using a recipe similar to Sutapa Roy Ramanan [14]. Briefly, indium chloride was dissolved in acetic acid and acetylacetone (1:0.5:0.5 molar ratio in solution), then an alcohol solution (isopropanol: 2-methoxyethanol: glycerol = 1:1:1 in volume ratio) was added to bring the overall solution to an In_2O_3 concentration of: 6% wt., 10% wt., and 20% wt. For each of these, a proper amount of stannic chloride was added afterwards, such that all three solutions maintained the In-Sn atomic ratios of 90:10. The ITO solutions were aged at least 1 h after stirring and then deposited onto the prepared optical fibers as ultra-thin layers. This was performed at room temperature with the fiber fully immersed in the sol–gel solution and then drawn out with a withdrawal speed of 8.0 cm/min to form an ultra-thin and uniform layer of approximately 40 nm thickness [15] using a PCTL0.01 DIP COATING system. After the deposition, the ITO-coated fibers were annealed in a long quartz tube (1.5 m) lying in a preheated furnace at 500°C. There are three temperature zones along the tube (1) furnace center that was preheated up to 500°C; (2) the edge of the entrance to the furnace (200°C ~ 300°C); (3) the entrance of the tube (~50°C). In the beginning of the annealing, we preheat the sample on zone (2) for 5 min and push it to zone (1); After 30 min annealing, we drag sample to zone(2) for 10 min and zone (3) for 10 min for cooling step by step [16–18]. After cooling, another layer was added using exactly the same protocol. Once a full set of multi-layers had been deposited, the resulting film was annealed in a N_2/H_2 (1%) atmosphere at 550°C for 1 h.

The linear resistivity was measured by a two-point probe method using a FLUKE 111 multi-meter with the probes placed at a distance of 1 cm along the fiber, for each of the films investigated. This yielded conductivity for the one-dimensional optical fibers, which clearly includes the contact resistance of the contacts. However, since this is presumed to be the same for each case, comparisons between fibers were used to determine the effects of layer number and conductivity. The thickness and grain structure were determined by FE-SEM imaging of fractured fiber cross-section.

5.1.1.2 Device Fabrication

To investigate the effect of these films on device performance, polymer-based photovoltaic cells were fabricated on the ITO coated fibers. First, the prepared ITO coated fibers were ultrasonically cleaned in the acetone and isopropyl alcohol bath for 5 min each and dried at 100°C. The fibers were exposed to ozone for 90 min. Poly(3,4-ethylenedioxythiophene)- poly(styrenesulfonate) (PEDOT:PSS) solution (Baytron P) was deposited by dip-coating and dried at 100°C for 15 min. Subsequently, a solution of P3HT: PCBM (1:0.8) in chlorobenzene was dip-coated onto the fiber yielding a film of approximately 200 nm. Finally, a 300 nm aluminum (Al) electrode was deposited using thermal evaporation. Devices were tested using an AM1.5 g standard (Oriel) operating with an illumination intensity of 100 mW/cm^2 applied at the cleaved end of the fibers. Current–voltage (J-V) characteristics were collected using a Keithley 236 source-measurement unit. It should be pointed out that the devices fabricated on the fibers were of the same length and placed at the same position relative to the cleaved (illuminated) end of the fiber. This was done to ensure that the "device" had the same area of absorber and that the same light intensity reached the "device" in each case. We note that if the devices are placed at different distances from the front face of the fiber, different amounts of light will be scattered out of the fiber before it reaches the photovoltaic. Further, the OPVs tested were not annealed, which yields a lower performing P3HT device generally, but avoids pinhole formation that sometimes occurs in fiber-based devices. Again, the intention is to investigate whether similar light transfer can be expected from the two cathode morphologies.

5.1.2 Performance of Fiber-Based Solar Cell

5.1.2.1 Film Morphology, Thickness and Linear Resistivity Measurement

Comparing the single layer ITO quality produced by the three solutions as described in Fig. 5.1, the best film was observed from the use of the 6% wt. of In_2O_3 solution mixed with $InCl_3$ and $SnCl_4{\cdot}5H_2O$ in alcohol. Generally speaking, the 6% wt. solutions produced a slightly thinner single layer than the higher concentration films (40 nm compared to 60 nm for the 20% wt.). Furthermore, the films appear continuous and smooth for this concentration. Naturally, it is expected that this will depend on the withdrawal rates and temperature of the beginning solution. However, generally, we found that the combination of withdrawal rate and room temperature worked best with the lower concentration.

Using the 6% wt. preparation, additional layers were added sequentially, following the protocols described above. The thickness of ITO film (as determined by FE-SEM on cleaved fibers) increases almost linearly with layer number as shown in Fig. 5.2a. The thickness of ITO film with 2 layers is around 50 nm. Optical microscopy suggests a smooth and continuous film. However, the scanning electron micrographs in Fig. 5.3d reveal that the films are not continuous or uniform at the scale of 10's of nanometers. As expected, it also shows a large linear resistance of 49.4 kΩ/cm. With increasing the layer number from 2 to 11, the thickness of the ITO film increases to 190 nm, while the ITO film resistance decreases to 1.3 kΩ/cm. The smooth and rapid fall off of resistivity indicates that the film is becoming more continuous and more conductive through the addition of more conducting material.

The thickness of ITO increases linearly with number of coatings, which is expected from other reported work. This yields the same trend in resistivity as a function of coating thickness [14]. The resulting sol–gel was quite stable. Typically

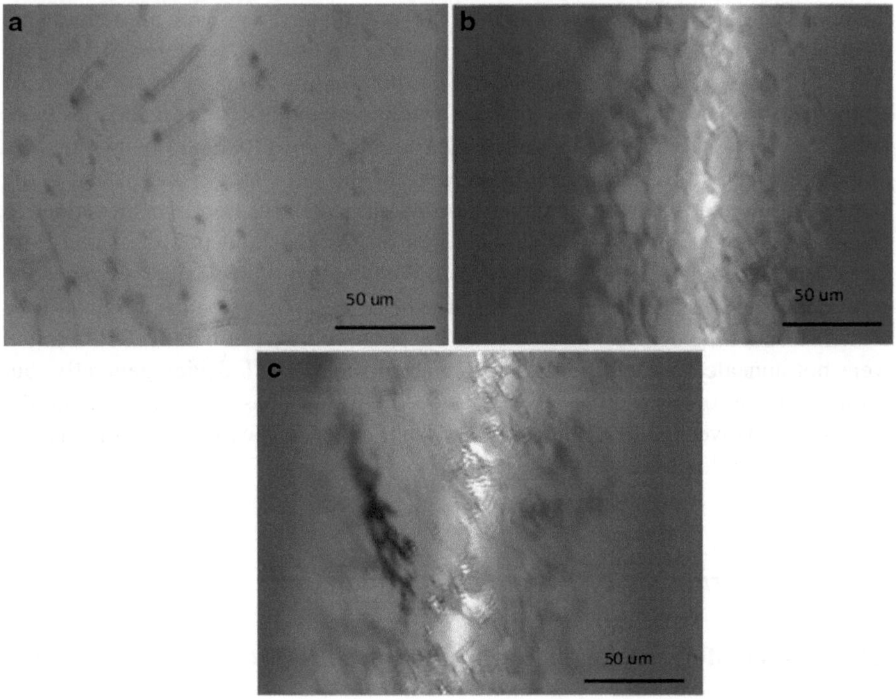

Fig. 5.1 (100×) Magnification optical micrographs of the fibers with one layer of ITO applied using solutions that contained (**a**) 6% wt (**b**)10% wt (**c**) 20% wt In_2O_3. These images are after drying

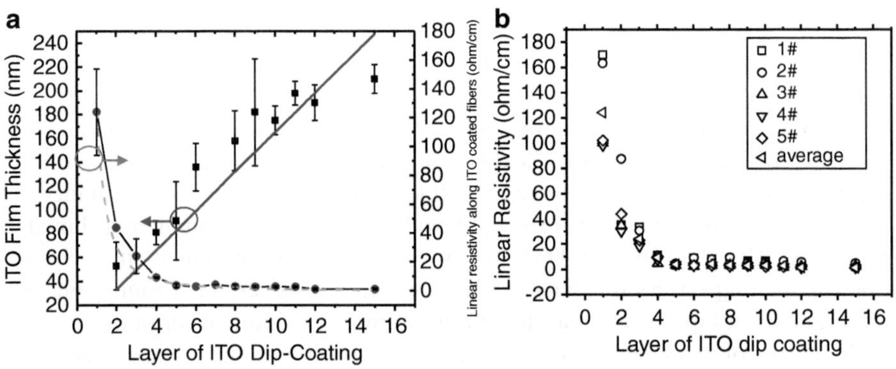

Fig. 5.2 (**a**) (*Left-red*) Film thickness as a function of the number of layers added. The concentrations for this example were 6% wt. Thickness of ITO film on fibers increases linearly with layer number; (*right-blue*) 2 probe, linear resistance along the ITO coated fiber decreases with number of layers. The *green line* is a fit to $1/n^2$, where n = layer number; (**b**) the statistical graph for linear resistivity (ohm/cm) for 5 fibers dip-coated from 6% wt ITO solution as a function of coating layers and the average value

one batch of fibers took 2–3 days to make and the three sample batches were made over different days. They all showed the same trends, as seen in Fig. 5.2. The average values are plotted in Fig. 5.2. The resistivity change can be attributed to the growth of the grain sizes [19]. As can be seen in the SEM images in Fig. 5.3 in the section, the grain grows from ~10 nm for one-layer coating to ~30 nm after four

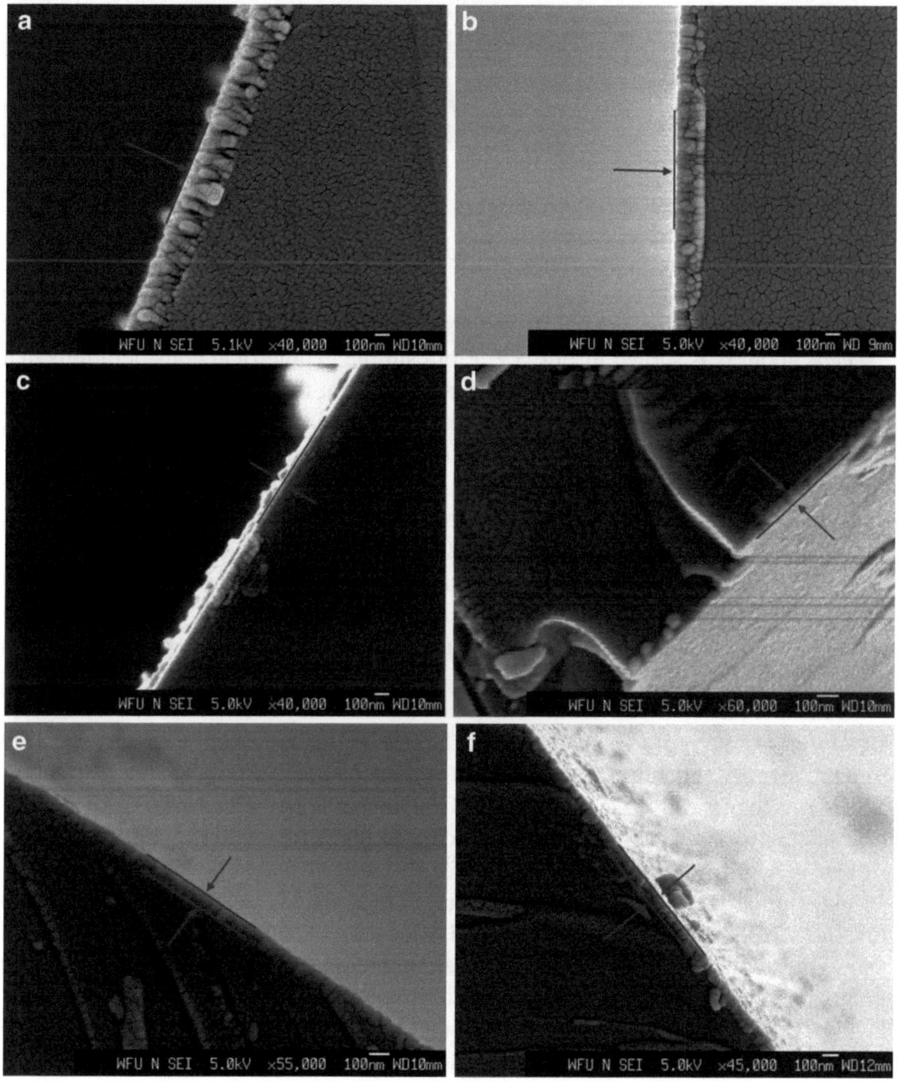

Fig. 5.3 (**a**) SEM of ITO film on 0.2 mm fiber after sputtering coating; SEM of ITO film on 0.2 mm fiber with (**b**) 11 layers (**c**)1 layer (**d**) 2 layers (**e**) 4 layers (**f**) 5 layers (**g**) 7 layers (**h**) 15 layers of the ITO solution. Fibers were prepared for imaging by fracturing and adding a thin layer of Au to reduce charging of the glass (the top, outer layer, of the fiber is marked in each micrograph)

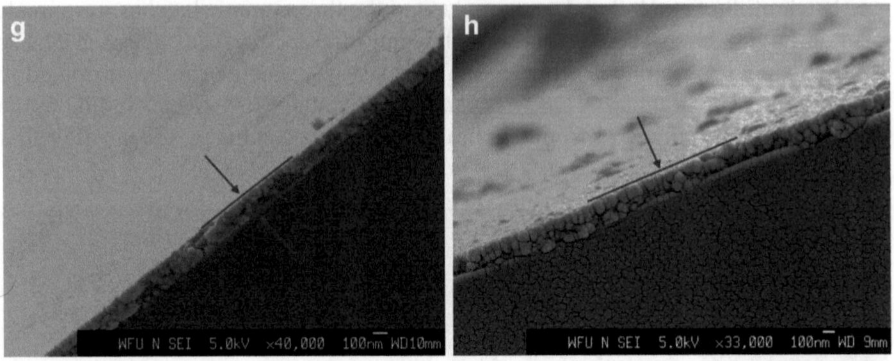

Fig. 5.3 (continued)

layer-coatings, and remains constant for the rest of the time. As noted, the linear resistance of the ITO coating decreases nearly exponentially with increasing layer number. This linear resistance then doesn't change much once the layer number exceeds about 6 as shown in Fig. 5.2. From simple geometrical considerations in a continuous multi-layer growth, we would expect that the conductivity would grow as layer number squared (n^2) and the resistivity would drop by this factor (n^{-2}). Fitting the resistivity above about six layers to n^{-2} where we are assured of a fully continuous film, we can see that even for layer numbers below 4 the increase in conductivity is consistent with continuous layer-by-layer growth. The best fit for conductivity was given by σ=124 (kΩ/cm)*n^{-2} and is shown by the green line in Fig. 5.2. Similar investigations were carried out on the other concentrations and on larger diameter fibers with similar results (not shown here). In comparison a sputter coated, single monolith ITO coating has a linear resistivity of 1.1 kΩ/cm at a thickness of 200 nm.

With increasing layer number, increasing smoothness and well defined grain structure is observed, as shown in Fig. 5.3. With 11 consecutive layers, the ITO film is generally very uniform and smooth around the fiber surface. Figure 5.3b shows a SEM image of the cross section of an ITO 11-layered fiber. The thickness of the ITO film is around 190 nm and compares nicely to that of a sputtered 200 nm film as shown in Fig. 5.3a. The overall grain size for the 11-layer film is approximately 30 nm, similar to that of the sputtered film. A reasonable range of 20 ~ 50 nm ITO thickness in planar films has been reported using sol–gel methods [20, 21]. Fig. 5.3c–h shows the evolution of the film as the layer number is increased. Notice that the ITO grains aggregate/grow from ~10 nm in the monolayer ITO to ~ 30 nm after four layers. Of course, the grain size is estimated in these micrographs from the overall diameter along the surface. We do note that the sputtered film grains exhibit a more columnar grain growth than that of the multi-layer film which has more rounded grain and with grains stacked one on another to achieve the overall thickness observed. This difference in morphology may explain the slight differences in 2-probe conductivity observed between the sputter coated and multi-layer films. It is also important to state that variations in stoichiometry can result in significant changes in the conductivity of such films [22]. To address this, both the multi-layer and sputter deposition coatings were post-treated in similar ways. Though this doesn't guarantee exactly the same stoichiometry between the films, we postulate that the overall stoichiometry of the grains is similar and the grain morphology is the primary driver between conductivity differences. Further experiments are planned to confirm this.

The optical transparency determination is hard to carry out even for a planar substrate due to the thickness dependence of each cycle of dip-coating as shown in Eq. 5.1 [23]:

$$h \propto \sqrt{\frac{\gamma}{\rho g}} Ca^{1/2} \tag{5.1}$$

Where, h, γ, ρg, Ca is thickness, surface tension, gravity and capillary number respectively.

Therefore, the thickness of dip-coated single layer not only depends on the viscosity of solution and velocity of withdrawing speed, but also relates to the surface tension which is a function of dimension/curvature of the object. So a flat substrate may not be able to reproduce the same characteristics film on optical fibers as was our worry.

5.1.2.2 OPV Performance on ITO Dip-Coated Fibers

Performance in a photonic or opto-electronic application depends not only on the electrical conductivity of the films but also on the optical properties such as transparency and scattering. It is natural to assume that the differences in granularity may lead to quite different behaviors in scattering and light-trapping near the surface of the optical waveguide, the fiber in this case. To understand how these films behave in a specific device, fiber photovoltaic devices were fabricated on the different multi-layer and sputter-deposited ITO fibers. These results can then be tested against an "ideal" ITO cathode through direct simulation of device performance.

Fabrication and testing of the photovoltaic devices is described above. A detailed structure of a fiber-based solar cell is shown in Fig. 5.4a. The J-V data collected from these types of devices, with an illumination angle of 30° [24, 25] are summarized in Table. 1 and J-V curves shown in Fig. 5.4b below. To calculate the effective efficiency of the devices, we use the following standard formula: $\eta_{\text{external}} = J_{sc} \cdot V_{oc} \cdot \text{FF}/P_{\text{incident}}$ where J_{sc} is the short circuit current density; V_{oc} is the open circuit voltage, FF is the filling factor calculated by taking the maximum power rectangle directly from the *J-V* curve, and P_{incident} is the intensity of the incoming light (100 mW/cm^2). The area of *normal* illumination is that of the fiber face and each fiber here is in length of ~1 cm for the active region.

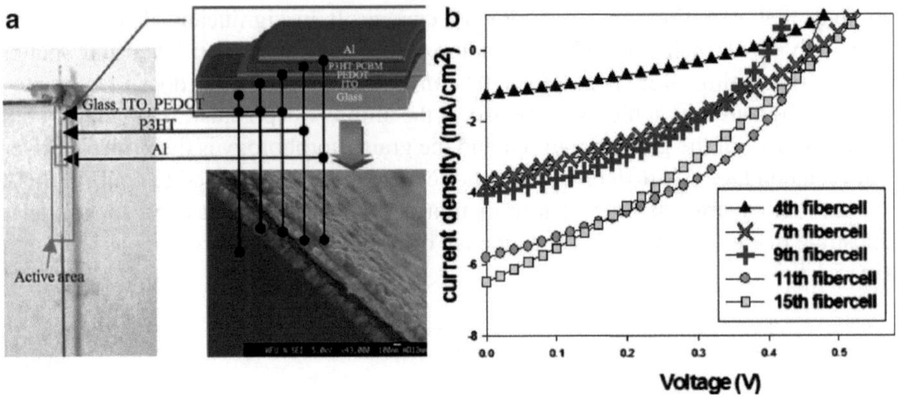

Fig. 5.4 (**a**) A detailed scheme illustration of fiber based solar cell; (**b**) IV curve under illumination for devices fabricated around 4-layer (*triangle*), 7-layer (*cross*), 9-layer (*plus*), 11-layer (*spot*) and 15-layer (*square*) dip-coated ITO fiber

It is clear that the lower resistance of the thicker films results in a higher efficiency as expected. The Jsc of the fiber solar cell with a nine layer ITO dip coating, 4.03 mA/cm^2 is much higher than that of the fiber device from four times ITO dip coating, 1.26 mA/cm^2, and the efficiency of the fiber device from nine layer ITO coating is about six times of that for the fiber device made from four layer ITO coating. The increase in filling factor with increasing number if layers in the multi-layer films indicates a decrease in internal resistance. This increase of FF occurs up to about 11 layers, but at 15 layers, near the thickness of the sputtered film, the filling factor drops. So while the overall current increases with the 15-layer device, its performance decreases due to the relatively high internal resistance. This is likely due to a morphology change in the thickest multi-layer deposited films. Since the grain morphology of the 15-layer films appears to be similar, this morphological change is associated with long range roughening as can be seen in the optical micrograph of Fig. 5.5.

5.1.2.3 Theoretical Simulations

Theoretical simulations were carried out to better determine the light transfer from the fiber to the absorbing layer as a function of layer number or thickness [26, 27]. The simulations use a ray tracing method coupled with a transfer matrix formalism which accounts for different materials as described in Chaps. 2, 3, and 4.

The detailed simulation methods are mentioned in reference [26, 27]. Several parameters are necessary in this model. The complex refractive indexes of ITO and glass were taken from reference [28]. The optical data for P3HT:PCBM were obtained from reference [29], and the optical data for Al were obtained from Refractive Index. Info. database (http://refractiveindex.info). The complex refractive index of PEDOT is obtained from reference [30]. Then, we used a transfer

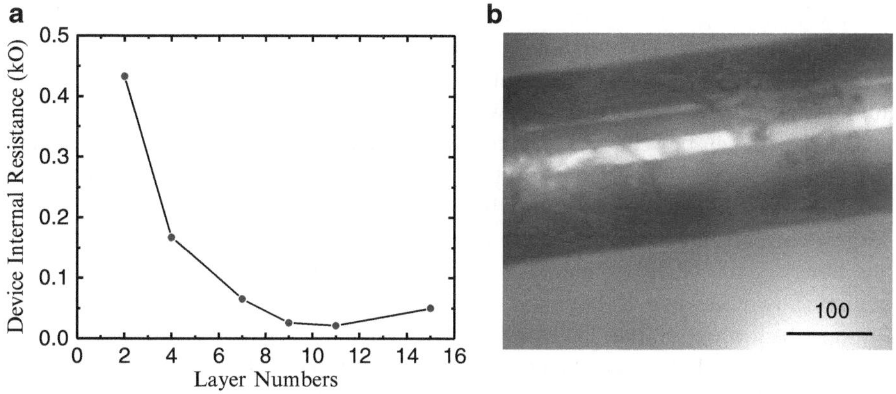

Fig. 5.5 (a) The internal resistance of the OPV devices as a function of ITO layer number; (b) (50×) magnification optical micrograph for the 15 layer ITO-fiber after thermal annealing showing regions of roughening

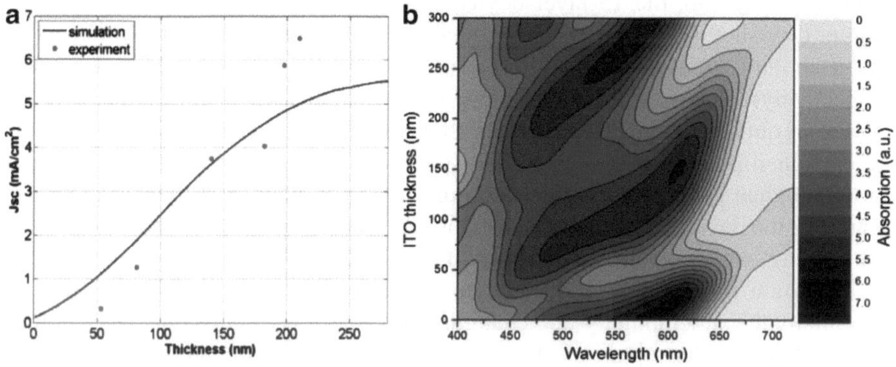

Fig. 5.6 Current density comparison between theoretical simulation and experimental data (a) With ITO thickness and wavelength varying, the absorption in P3HT:PCBM is calculated using a transfer matrix model of fiber cell. This absorption and equivalent circuit of fiber cell are then used to simulate the short circuit current in (b)

matrix method based on OPVAP [31] to simulate the optical field distribution. We should note that the thickness of ITO is defined as 100 nm, PEDOT is defined as 40 nm, P3HT:PCBM is defined as 130 nm and Al is defined as 100 nm.

The experimental data (red dots) are also plotted on the theoretical curve in Fig. 5.6a. Notice that the theoretical current density gets stronger sharply from 1.26 mA/cm^2 for 81 nm ITO to 5.0 mA/cm^2 at 180 nm where it becomes relatively stable. This trend is followed closely by the experimental data. This suggests that a maximum amount of light is being transferred into the absorbing layer through the film corresponding almost exactly with sputter deposited films of 6.21 mA/cm^2 at a thickness of 220 nm.

It's interesting and important to note that the performance of the fiber device doesn't improve as ITO thickness (number of layers) is increased beyond the 11 layers as shown above. To explain this, we performed simulations to better understand how the optical power was distributed within the absorber and in the ITO [31]. As can be seen in Fig. 5.6b, the theoretical calculations show that there are multiple peaks of energy absorption in the P3HT layer with the change of ITO thickness as a function of wavelength. The largest number of photons, and thus the largest currents, are expected where the loss to the ITO layer is smallest, shown in the blue regions of the graph. The rapid fall off of absorption in the photoactive region shown between 400 nm and 450 nm for all reasonable thicknesses is due to strong absorption in the ITO and PEDOT layers. However, large absorption features in the photoactive layer occur at around 50–100 nm, 120–200 nm and 250 nm in thickness. These variations are due entirely to optical effects within the device. From the above, the conductivity is severely limited in the 50–100 nm films, so the best performance is expected to occur for film thicknesses in the region of 120–150 nm and between 200 and 220 nm since these regions yield the largest integrated absorption across the wavelength range. The 11 layer device with ITO thickness ~198 nm, the 15 layer device with ITO thickness of ~210 nm and the sputter deposited ITO device (~220) are all relatively within this range, suggesting that a similar current should be observed (5.88, 6.49, and 6.21 mA/cm^2 are observed respectively). However, the internal resistance as indicated by filling factor of the devices is quite different. Thus, the variation seen in the 15 layer ITO device cannot be explained by optical effects alone. Again, as postulated above, this suggests a morphological change in the thickest multi-layer device which affects the performance of the device. But this variation must be electronic in nature such as surface roughening, which would lead to poor filling factors.

The grain morphology and absorption characteristics of the ITO, as well as potentially the stoichiometry, are changing rather strongly in this thickness regime. We attribute the discrepancies at this point of the curve to the model's inability to account for such changes.

As stated briefly above, the efficiency of the devices studied here is limited by the position of the device along the fiber. In order to prevent the devices from shorting out and to make them easy to fabricate, the fiber OPV device is actually built approximately 4 mm from the fiber end. The ITO extends out from the device to allow contact with leads. This means that a significant amount of light from the AM1.5 g diffuse light standard does not couple properly into the fiber and that light is lost without producing any current. This means that the current density observed in these experiments is considerably lower than what is ultimately possible for devices fabricated near the fiber end. Every attempt was made to place the device at exactly the same position along the fiber for each of the devices studies. However, the error in placement is enough to lead to the variations between the currents measured from the 11-layer, 15-layer and sputtered devices. This error would not lead to such variations in the filling factor. Again, this supports our conclusion that the 15 layer device has experienced a change in its electronic behavior, leading to increased internal resistance.

In sum, this section introduced a method for the creation of high quality, conformal ITO films using a consecutive multi-layer process. Resistivity and grain morphology of the films are similar to ITO films prepared by sputtering methods but provide excellent coating to three dimensional surfaces. The resistivity of the multi-layer coated ITO film decreases as the number of layers increases. Further, we have shown that the performance of fiber-based photovoltaic devices prepared using multi-layer ITO coatings are comparable to that of fiber OPV devices made from sputtered ITO films. This suggests that light delivery through the multi-layer ITO film is also quite similar to that of sputtered ITO, but there are limits. Finally, we submit that multi-layer ITO deposition is an efficient way to prepare transparent conductive cathodes for shaped organic devices.

5.2 ITO-Free Tube Based Solar Cell

A wave-guiding tube-based optical geometry provides a general approach to improving organic photovoltaic performance in Fig. 2.4b. By fabricating bulk-heterojunction photovoltaics onto thin tubular light pipes, the optical energy can be effectively captured within the absorbing layer without associated reflective losses at the front and rear surfaces of the devices, as is typical in planar structures. We have used a common absorber system: poly-3-hexyl-thiophene:phenyl-C61-butyric-acid-methyl-ester to demonstrate that these tubular optical geometries result in very little overall radiative loss. Surprisingly, this also leads to an overall broadening of the absorption window for the device as indicated by the external quantum efficiency (Reprinted with permission from [32], Copyright (2010), American Institute of Physics).

In the past 10 years, several groups have reported advances in the synthesis of small-band-gap conjugated polymers [33–36] that promise to significantly broaden the spectral overlap of organic photovoltaics absorbers. Poly(5,7-bis(4-decanyl-2-thienyl)-thieno(3,4-b)diathiazole-thiophene-2,5) (PDDTT), for example, has a spectral response from 300 to 1,450 nm and exceptional internal quantum conversion efficiency [37]. These low band gap polymers have already demonstrated efficiencies in planar devices exceeding 6.7% [38–40]. However, if one determines the total number of photons in the spectrum between 300 and 1,450 nm and then assumes each photon could result in one electron–hole pair, then efficiencies much higher than this should be expected. Indeed, efficiencies greater than 14% should be easily achieved. So why isn't this the case? A significant drawback to the use of any polymer absorber is that reflections at the surfaces and interference effects within the thin film prevent light from being absorbed. What is needed is an optical geometry that allows for all of the light within the absorption band to be captured.

This optical geometry can be achieved by 'wrapping' the solar cell around a tubular cavity. Similar in design to the fiber-based photovoltaics which have recently been reported [4–6, 13, 24], tube-based photovoltaics (TBPV) have no loss by reflection at the front surface. Further, due to the special "tube structure",

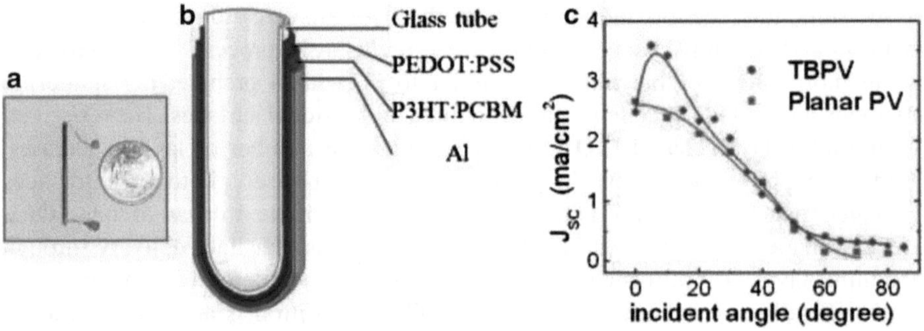

Fig. 5.7 (**a**) Is the Schematic of the TBPV architecture and light illumination (longitudinal section) and (**b**) is a photographic image of the TBPV device. Since the closed end face is silvered by Al, the light will reflect on this curved end face and continue to transmit back up the tube to be absorbed. (**c**) Comparison of the Jsc for TBPV and planar photovoltaic as a function of the incident angle

light will be continuously transmitted down the tube and be completely absorbed. We note that this is quite different from absorbing light along the outer surface of a large cylindrical tube as in the case of Solyndra's devices [41]. While we use thin films of Poly 3-hexyl thiophene:phenyl-C_{61}-butyric acid methyl ester (P3HT: PCBM) to demonstrate the function of TBPVs, the light confinement principles can be applied to any absorber. Thus, it represents a general approach to dealing with the thin-film effects in organic photovoltaics.

TBPV devices were fabricated on glass tubes with one end closed in a hemispherical cap (Chemglass, 0.8 mm inner diameter, 0.25 mm wall, 1.3 mm outer diameter). The tubes were cleaned in an ultrasonic bath and were then exposed to ozone for 90 min (rotating the tubes three times every 30 min). Subsequently, by a dip coating procedure (Opticer H10 Dip Coater), poly(3,4-ethylenedioxythiophene)-poly(styrenesulfonate) (PEDOT:PSS, Clevios^{TM} PH 750, conductive, thickness ~200 nm), P3HT:PCBM(1:0.8 in weight, of 12 mg/ml in chlorobenzene, thickness ~80 nm) were deposited on the tubes. Conductive PEDOT was used in place of the standard indium tin oxide (ITO) due to ease of processing (dip-coating) and quality of the final film. Finally, Al electrodes were deposited via thermal evaporation at a pressure of 10^{-6} torr. Contacts were silver pasted onto the tube. The length of the tube with active area was about 4 cm. The resulting device structure is shown in Fig. 5.7.

TBPV devices were tested using an AM1.5 g standard Newport #96000 Solar Simulator with an illumination intensity of 100 mW/cm^2. The open end of the tubes was illuminated. Current voltage characteristics were collected using Keithley 236 source-measurement unit. The transmittance spectra of all films were characterized by an UV–visible is spectrophotometer Varian, Cary 50. The external quantum efficiency (EQE) was measured using a Newport Cornerstone 260 Monochromator in conjunction with a Newport 300 W Xenon light source. Apertures ensured that side illumination did not occur. Planar photovoltaics were fabricated on glass using the usual spin casting techniques as a control, and tested used the same configuration of polymers with no ITO to ensure an accurate comparison.

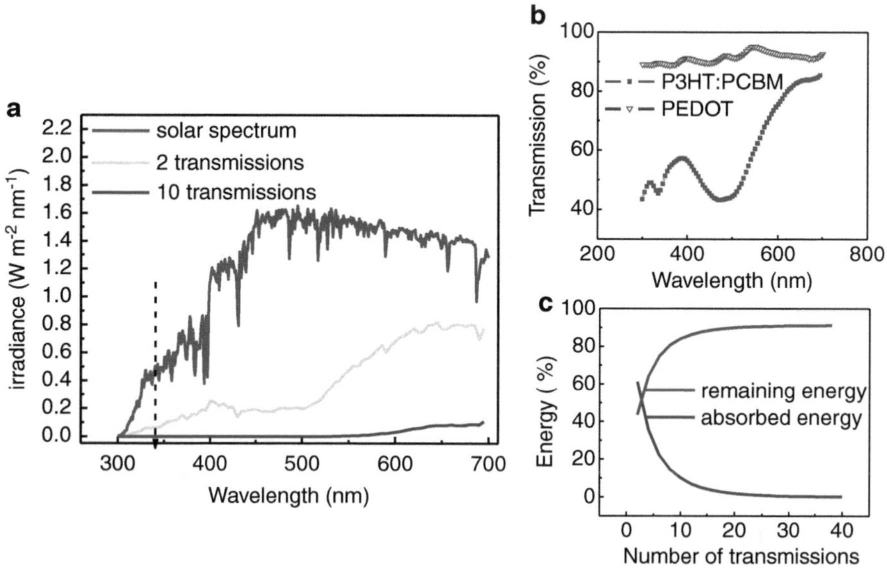

Fig. 5.8 (**a**) A comparison of the solar spectrum as it transmits through the layers of the two photovoltaics. The curve (2 transmissions) corresponds to the solar energy remaining after a single reflection through a planar device. The curve (10 transmissions) represents the remaining solar energy after the light is transmitted down a TBPV once. (**b**) The transmission spectrum of P3HT: PCBM, and PEDOT used to compute the curves of (**a, c**) shows that the energy absorbed by multiple transmissions through the P3HT:PCBM layer (*bottom line*) in the TBPV grows quickly as a function of internal reflections in the device. The remaining light energy in tube (*top line*), shows the corresponding decreases to zero

Using the transmission spectrum (Fig. 5.8b) of each material with thicknesses of device and an iterative approach (Fig. 5.8a and Eq. 5.3), we can estimate how much energy will be absorbed by continuous transmission and absorption in a TBPV comparing with planar PV in (a). As would be expected, significantly more light than that for the planar device should be absorbed. In terms of this method, there is 35% energy that is not absorbed in thin film planar PV in Fig. 5.8c, but is utilized by TBPV. If we assume that the TBPV and the planar photovoltaic can be made with the same overall voltages and filling factors then the efficiencies of the devices, defined as the open circuit voltage (Voc) times the filling factor (FF) times the short circuit current density (Jsc) divided by the input solar power: (Voc $\times$ FF $\times$ Jsc)/Solar Power, can be divided such that the ratio of efficiencies between the two devices is simply the ratio of their absorbed energies. This too assumes that the conversion of absorbed photons into current is the same for both cases. In fact this yields an underestimate of the advantages of TBPV since photons emitted by radiative recombination can have a greater chance of being absorbed due to the confined nature of the tube geometry as opposed to the open nature of the planar photovoltaic. Neglecting this correction, the efficiency of TBPV could be obtained by:

$$
\begin{aligned}
\text{Ratio}_{\text{TBPV/Planar PV}} &= \frac{\text{efficiency of TBPV}}{\text{efficiency of Planar PV}} \\
&= \frac{\text{absorbed energy of TBPV/unit area}}{\text{absorbed energy of Planar PV/unit area}} \\
&= \frac{\lim\limits_{n\to\infty}\{a_{P3HT:PCBM}\cdot S + a_{P3HT:PCBM}\cdot(1-\mathrm{A})\cdot S + \cdots + a_{P3HT:PCBM}\cdot(1-\mathrm{A})^n\cdot S\}}{\mathrm{a}_{P3HT:PCBM}\cdot S + \mathrm{a}_{P3HT:PCBM}\cdot(1-\mathrm{A})\cdot S} \\
&= \frac{\lim\limits_{n\to\infty}\{1+(1-\mathrm{A})+(1-\mathrm{A})\cdot(1-\mathrm{A})+\cdots+(1-\mathrm{A})^n\}}{2-A}\,\text{(Neumann series)} \\
&= \frac{1}{A(2-A)}
\end{aligned}
\tag{5.2}
$$

The total absorption spectrum $A = \Sigma a_i$, where a_i is an absorption spectrum (expressed as a vector) of material i. For example, $a_{P3HT:PCBM}$ is the absorption of layer P3HT:PCBM; S is the solar spectrum; and n is the transmission time. Thus, the efficiency of TBPV equals to the product of the efficiency of planar photovoltaic and $\text{ratio}_{\text{TBPV/Planar Photovoltaic}}$ by Eq. 5.3:

$$
f_{\text{efficiency of TBPV}} = \frac{\sum\limits_{\text{each wavelength}} (A(2-A))^{-1}S}{\sum\limits_{\text{each wavelength}} S} f_{\text{efficiency of Planar PV}}(A) \tag{5.3}
$$

Where, $f_{efficiency\ of\ TBPV}$ is the estimated maximum efficiency for TBPV in terms of the efficiency of planar photovoltaic that uses the same materials as the TBPV.

Using Eq. 5.3, and the efficiency of planar photovoltaic with the same layers ($\eta_e = 0.52\%$; in Fig. 5.9b), the efficiency of the equivalent TBPV can be calculated to be $\eta_{e\_max} = 0.80\%$. Keep in mind this is a relative efficiency based upon PEDOT/PCBM:P3HT absorbers on glass using no ITO and with relatively low filling factors and no annealing. The improvement in the performance due to changes in the optical architecture is striking.

An important and significant prediction that can be made from this analysis is that the TBPV should absorb far more light at the band edges of the polymer. In the case of P3HT:PCBM systems, this increased absorption will occur between 600 and 700 nm. Typically, these states would be filled with relatively non-mobile charge, and thus not contribute much to the overall power generation. To see if this is the case, we measured the External Quantum Efficiency of the TBPV and planar photovoltaic devices using the same materials in Fig. 5.9a. Across most of the absorption spectrum we see enhanced power generation for the TBPV as should be expected from its greater ability to capture and retain the solar flux. The improvement of current generated by EQE from 300 to 800 nm is about 30% relative to planar PV. We also see that power generation occurs at longer wavelengths for the

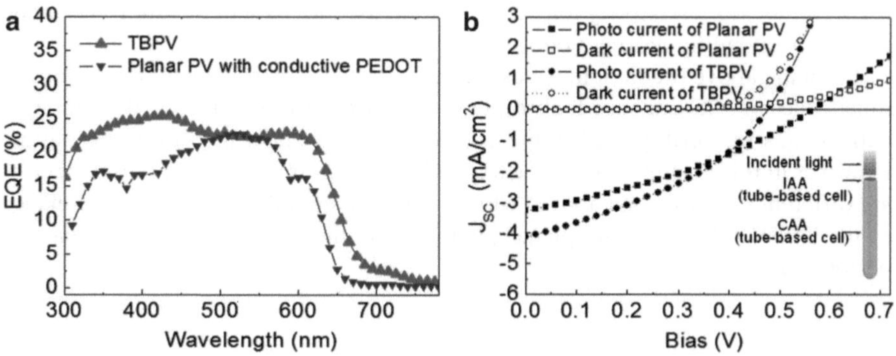

Fig. 5.9 (**a**) EQE versus wavelength of the two photovoltaics: (*Down triangle*) planar photovoltaic with conductive PEDOT/P3HT:PCBM/Al; (*Upper triangle*) TBPV with conductive PEDOT/P3HT:PCBM/Al. The improvement of current generated by this EQE is about 30% relative to planar PV. (its corresponding current density changes from 3.1 to 4.1 mA/cm^2) (b) J-V characteristics of TBPV and planar photovoltaic (*center*), and active area explanations (*bottom right corner*). The current density of short circuit is defined as $J_{sc} = I_{sc}/A_{ea}$, where A_{ea} is the IAA. This is a device current density. The power conversion efficiencies of the solar cells are given as: **TBPV** $J_{sc} = 3.5$ mA/cm^2, $V_{oc} = 0.475$ V, FF $= 0.37$, and $\eta_e = 0.60\%$; **planar photovoltaic** $J_{sc} = 2.8$ mA/cm^2, $V_{oc} = 0.565$ V, FF $= 0.34$, and $\eta_e = 0.52\%$. Current improvement is 25%, and efficiency improvement is 15%

tubular devices than for the planar devices. This suggests that the band edge absorption in the TBPV is capable of generating current (and thus power), a quite unexpected result.

With the broadening of the EQE spectrum in the TBPV, as shown in Fig. 5.9a, a significant increase in overall device performance is expected. Shown in Fig. 5.9b (the light and dark curves of the TBPV as compared to the planar photovoltaic.), the current improvement is 25%, and the efficiency improvement is 15%. To understand exactly how these were determined, we must introduce definitions for the two kinds of "active areas" in the TBPV: the "current active area" (CAA) and "illumination active area" (IAA). In Fig. 5.9b, the solar light enters the tube from the top open end face (IAA $= \pi$ O.D.$^2/4 \approx 0.01$ cm^2) and is absorbed on the side face of tube (CAA $= \pi$ O.D. tubelength $+ 2\ \pi \bullet (\text{O.D.}/2)^2 = 1.66$ cm^2). In addition, if the tubes are bundled together, it is necessary to take into account the area of gap between each tubes. Thus, $\text{CAA}_{\text{all}} = \text{n} \times \text{CAA}_{\text{single}} + \text{n} \times \text{gap}$, and IAA $=$ area of substrate supporting bundles. In sum, CAA is a concept we defined for 3D structure PV, such as tube-based solar cell, fiber-based solar cell, fiber bundles, etc [6]. Therefore, the currents in a TBPV can be high relative to the planar photovoltaic because the current generated by CAA is divided by IAA. This is done in the same fashion as previously reported for fiber-based photovoltaics [24]. Of course it must be stated that making the tube too long results in all of the light being lost in the first few millimeters of the device, leaving the back of the device to produce no current. Consequently, the Voc is lowered for devices that are too long. (decreases

0.1 V) In our work relatively long devices were used for ease of handling and we do see a corresponding drop in voltage.

From the above prediction, we expect an efficiency improvement of roughly 1.5 times due to the enhanced absorption properties of the TBPV device. This corresponds reasonably well to the 1.3 times improvement in current generation. However in this example the fill factor, and the open circuit voltage of two devices are different (as explained above) which reduces the total increase in performance of the TBPV relative to the planar photovoltaic, though a significant increase in performance is seen.

5.3 Tandem OCGPV

5.3.1 Serial Connection

Tandem architecture, using several sub-cells absorbing light piecewise, is applied on individual fiber and tube to further expand light harvest. Like planar tandem cells have, tandem fiber-based solar cells also have serial and parallel connections. However, a tandem fiber cell shows some interesting performances different from a planar cell, such as simple layer design and special connection style. In 2009, we first tried a serial tandem fiber cell based on P3HT:PCBM BHJ system as shown in Fig. 5.10 (left), which is proposed to improve Voc. Devices were built on individual fibers by the dip-coating method mentioned above. The ITO coating was etched in two parts, and the polymer layers were deposited on each part by dip coating.

The Voc of the tandem fiber cell was improved to 0.9 V, which is almost double that of a normal fiber cell device. However, device's Jsc is 2.2 mA/cm^2, because the

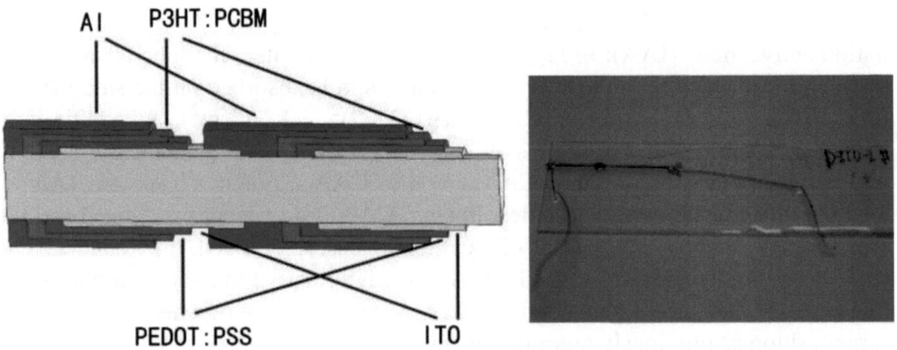

Fig. 5.10 The Schematic of the tandem fiber cell architecture (*left*) and corresponding picture (*right*)

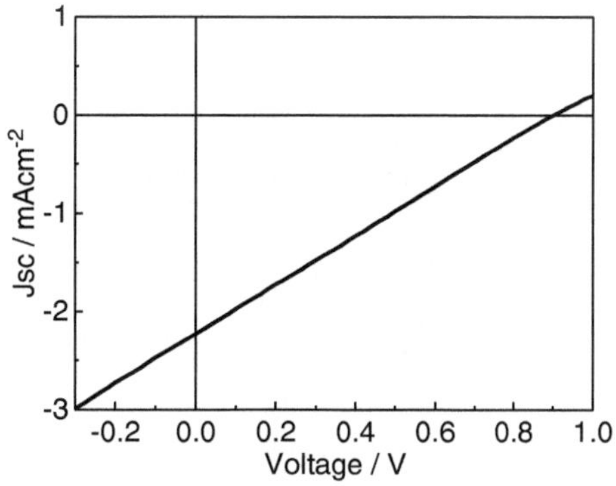

Fig. 5.11 Illuminated IV curve from a tandem fiber cell device

serial connection leads to a Jsc drop off due to the limit between current sources (proved in Sect. 1.2.4.1.2), and because there is a small gap on the Al coating in the center part of the device from where light is lost. Furthermore, the FF is only 0.25 which is explained by the rough film fabricated by the earlier dip-coating procedure. The resulting efficiency of 0.48% was achieved in Fig. 5.11. Two polymer layers with different absorption bands will be used on two parts of the tandem fiber cell in our further work. Then, a more effective tandem fiber cell should be achieved, as expected.

5.3.2 Parallel Connection

All planar tandem connections, whether serial or parallel, include extra layers [6], such as Ca, Ag, TiOx between sub-cells. Therefore, it is inevitable to have extra light loss from absorption by these inserted layers. However, by means of OCGPV, it is possible to make a parallel connection without inserting any extra layer, as shown in Fig. 5.12.

Achieving high efficiency requires balancing Voc and Jsc to reach an optimal product, for which we designed the architecture and chose polymers for parallel tandem OCGPV. Since OCGPV has the ability to fully transfer light to current in the range of the absorption spectrum, a small band gap material poly[2,6-(4,4-bis-(2-ethylhexyl)- 4H - cyclopenta[2,1-b;3,4-b1] dithiophene)- alt-4,7- (2,1,3-benzothiadiazole)] (PCPDTBT or ZZ50) is chosen to be the main contributor to Jsc (Fig. 5.13 shows its wider EQE in a tube-base cell). In addition, another polymer PBDTTPD is used to offset the voltage loss in OCGPV, because it has high Voc more than 0.8 V [42–45], which was synthesized by the Stille coupling polymerization between 1,3-dibromo-5-octylthieno[3,4-c] pyrrole-4,6-dione and di(2-ethylhexyloxy) benzodithiophene distannane monomers [44]. Chemical structures of these two polymers are shown in Fig. 5.13.

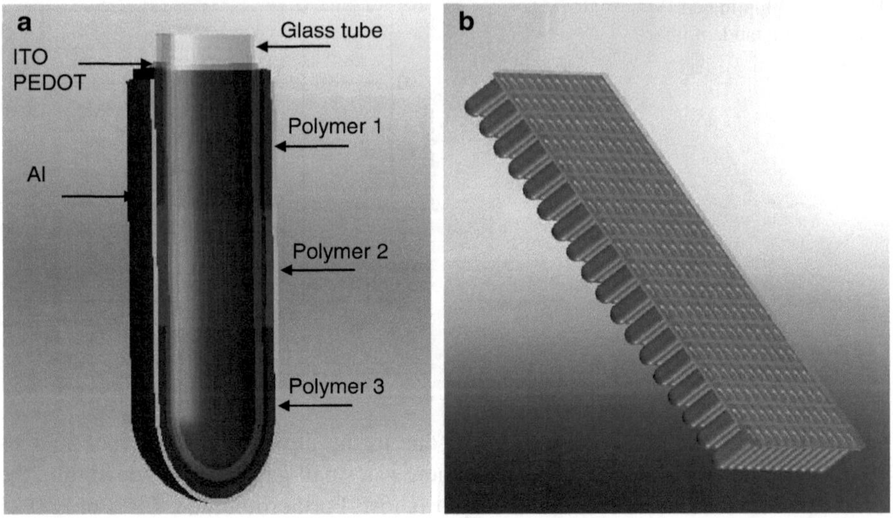

Fig. 5.12 (**a**) Architecture of parallel tandem solar cell. Polymer 1, 2 and 3 are represented as *red*, *blue*, and *dark green*, respectively. (**b**) Industrial parallel solar cell with large panel

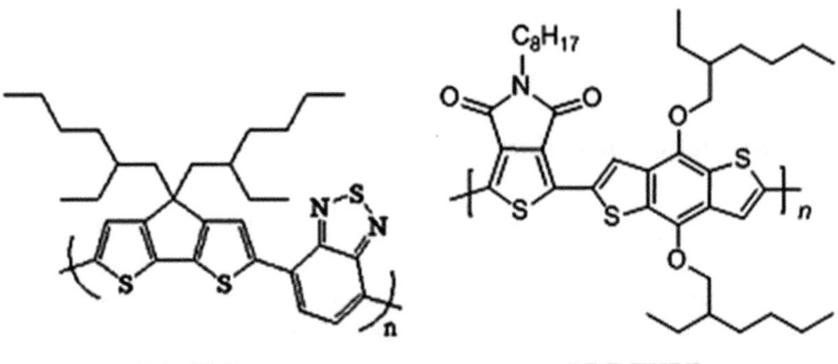

Fig. 5.13 Chemical structures of PCPDTBT (ZZ50) and PBDTTPD

As shown in Fig. 5.14 (left), the tandem OCGPV shows Jsc = 8.1 mA/cm^2 that is more than the sum of two single cells. This is explained as the absorption improvement by OCGPV, which was proved in the EQE spectrum of Fig. 5.14 (right). Furthermore, the Voc of parallel OCGPV equals to a value between Voc of two sub-cells, instead of choosing the minimal voltage as Voc, as proved in Sect. 1.2.4.1.2. The efficiency of tandem OCGPV is improved by nearly 100% from 0.21% (or 0.22%) to 0.44%, although the absolute value is not high due to the problems of device fabrication.

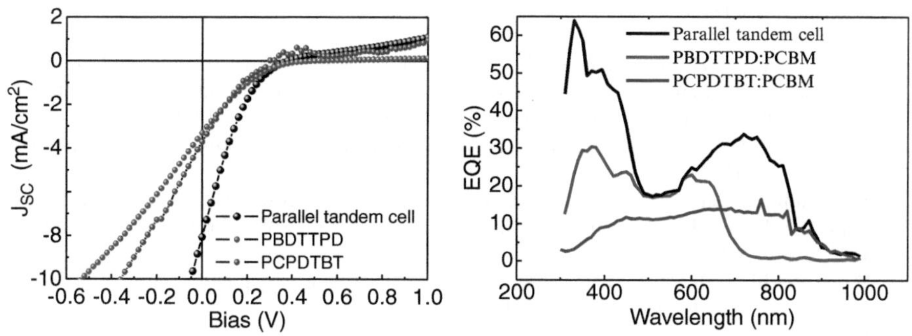

Fig. 5.14 (*Left*) J-V characteristics of two single cell (*red* and *blue*) and parallel cell with both polymers (*black*). The PCPDTBT:PCBM single cell shows J_{sc} =3.3 mA/cm^2, V_{oc} = 0.48 V, FF = 0.14, and η_e = 0.22%; The PBDTTPD:PCBM single cell shows J_{sc} =3.7 mA/cm^2, V_{oc} = 0.30 V, FF = 0.19, and η_e = 0.21%; and The tandem cell shows J_{sc} =8.1 mA/cm^2, V_{oc} = 0.38 V, FF = 0.15, and η_e = 0.44%. (*Right*) EQE of tube-based solar cell with each material and their parallel tandem cell

References

1. Y. Li, W. Zhou, D. Xue, J. Liu, E. Peterson, W. Nie, D. Carroll, Origins of performance in fiber-based organic photovoltaics. Appl. Phys. Lett. **95**, 203503
2. B. O'Connor, K. An, Y. Zhao, K. Pipe, M. Shtein, Fiber shaped light emitting device. Adv. Mater. **19**, 3897–3900 (2007)
3. B. Weintraub, Y.G. Wei, Z.L. Wang, Optical fiber/nanowire hybrid structures for efficient three-dimensional dye-sensitized solar cells. Angew. Chem. Int. Edit. **48**, 8981–8985 (2009)
4. J.W. Liu, M.A.G. Namboothiry, D.L. Carroll, Fiber-based architectures for organic photovoltaics. Appl. Phys. Lett. **90**, 063501 (2007)
5. J.W. Liu, M.A.G. Namboothiry, D.L. Carroll, Optical geometries for fiber-based organic photovoltaics. Appl. Phys. Lett. **90**, 133515 (2007)
6. Y. Li, W. Nie, J. Liu, A. Partridge, D.L. Carroll, The optics of organic photovoltaics: Fiber-based devices. IEEE J. Sel. Top. Quant. Electron. **16**, 1827–1837 (2010)
7. S.K. Park, J.I. Han, W.K. Kim, M.G. Kwak, Deposition of indium-tin-oxide films on polymer substrates for application in plastic-based flat panel displays. Thin Solid Films **397**, 49–55 (2001)
8. H. Kim, J.S. Horwitz, G.P. Kushto, Z.H. Kafafi, D.B. Chrisey, Indium tin oxide thin films grown on flexible plastic substrates by pulsed-laser deposition for organic light-emitting diodes. Appl. Phys. Lett. **79**, 284–286 (2001)
9. C. May, J. Strumpfel, ITO coating by reactive magnetron sputtering-comparison of properties from DC and MF processing. Thin Solid Films **351**, 48–52 (1999)
10. A. Salehi, The effects of deposition rate and substrate temperature of ITO thin films on electrical and optical properties. Thin Solid Films **324**, 214–218 (1998)
11. A.B.. Chebotareva, G.G. Untila, T.N. Kost, S. Jorgensen, A.G. Ulyashin, ITO deposited by pyrosol for photovoltaic applications. Thin Solid Films **515**, 8505–8510 (2007)
12. H. Mbarek, M. Saadoun, B. Bessais, Screen-printed Tin-doped indium oxide (ITO) films for NH_3 gas sensing. Mat. Sci. Eng. C-Bio. S. **26**, 500–504 (2006)
13. B. O'Connor, K.P. Pipe, M. Shtein, Fiber based organic photovoltaic devices. Appl. Phys. Lett. **92**, 193306 (2008)

14. S.R. Ramanan, Dip coated ITO thin-films through sol–gel process using metal salts. Thin Solid Films **389**, 207–212 (2001)
15. C.J. Brinker, A.J. Hurd, Fundamentals of sol–gel dip-coating. J. Phys. III **4**, 1231–1242 (1994)
16. M.J. Alam, D.C. Cameron, Investigation of annealing effects on sol–gel deposited indium tin oxide thin films in different atmospheres. Thin Solid Films **420**, 76–82 (2002)
17. C.J. Brinker, G.C. Frye, A.J. Hurd, C.S. Ashley, Fundamentals of sol–gel dip coating. Thin Solid Films **201**, 97–108 (1991)
18. K. Daoudi, B. Canut, M.G. Blanchin, C.S. Sandu, V.S. Teodorescu, J.A. Roger, Densification of In2O3: Sn multilayered films elaborated by the dip-coating sol–gel route. Thin Solid Films **445**, 20–25 (2003)
19. Y. Takahashi, S. Okada, R.B.H. Tahar, K. Nakano, T. Ban, Y. Ohya, Dip-coating of ITO films. J. Non-Cryst. Solids **218**, 129–134 (1997)
20. J. Herrero, C. Guillen, Improved ITO thin films for photovoltaic applications with a thin ZnO layer by sputtering. Thin Solid Films **451–52**, 630–633 (2004)
21. H.Y. Valencia, L.C. Moreno, A.M. Ardila, Structural, electrical and optical analysis of ITO thin films prepared by sol–gel. Microelectr. J. **39**, 1356–1357 (2008)
22. M. Mizuhashi, Electrical-properties of post-oxidized In2o3-Sn films. Thin Solid Films **76**, 97–105 (1981)
23. J.H. Snoeijer, G. Delon, M. Fermigier, B. Andreotti, Avoided critical behavior in dynamically forced wetting. Phys. Rev. Lett. **96**, 174504 (2006)
24. Y. Li, W. Zhou, D. Xue, J. Liu, E.D. Peterson, W. Nie, D.L. Carroll, Origins of performance in fiber-based organic photovoltaics. Appl. Phys. Lett. **95**, 203503 (2009)
25. Y. Li, E.D. Peterson, H. Huang, M. Wang, D. Xue, W. Nie, W. Zhou, D.L. Carroll, Tube-based geometries for organic photovoltaics. Appl. Phys. Lett. **96**, 243505 (2010)
26. Y. Li, H. Huang, M. Wang, W. Nie, W. Huang, G. Fang, D.L. Carroll, Spectral response of fiber-based photovoltaics. Sol. Energ. Mat. Sol. C. **98**, 273–276 (2012)
27. Y. Li, M. Wang, H. Huang, W. Nie, Q. Li, E.D. Peterson, R. Coffin, G. Fang, D.L. Carroll, Influence on open-circuit voltage by optical heterogeneity in three-dimensional organic photovoltaics. Phys. Rev. B. **84**, 085206 (2011)
28. H. Hoppe, N.S. Sariciftci, D. Meissner, Optical constants of conjugated polymer/fullerene based bulk-heterojunction organic solar cells. Mol. Cryst. Liq. Cryst. **385**, 233–239 (2002)
29. F. Monestier, J.J. Simon, P. Torchio, L. Escoubas, F. Florya, S. Bailly, R. de Bettignies, S. Guillerez, C. Defranoux, Modeling the short-circuit current density of polymer solar cells based on P3HT: PCBM blend. Sol. Energ. Mat. Sol. C. **91**, 405–410 (2007)
30. M. Caglar, S. Ilican, Y. Caglar, F. Yakuphanoglu, The effects of Al doping on the optical constants of ZnO thin films prepared by spray pyrolysis method. J. Mater. Sci-Mater El. **19**, 704–708 (2008)
31. Open Photovoltaics Analysis Platform (OPVAP) by Yuan Li, USA, www.OPVAP.com
32. Y. Li, E. D. Peterson, H. Huang, M. Wang, D. Xue, W. Nie, W. Zhou, and D. L. Carroll, Tube-based geometries for organic photovoltaics. Appl. Phys. Lett. **96**, 243505
33. S.H. Park, A. Roy, S. Beaupre, S. Cho, N. Coates, J.S. Moon, D. Moses, M. Leclerc, K. Lee, A. J. Heeger, Bulk heterojunction solar cells with internal quantum efficiency approaching 100%. Nat. Photonics **3**, 297–U295 (2009)
34. H.Y. Chen, J.H. Hou, S.Q. Zhang, Y.Y. Liang, G.W. Yang, Y. Yang, L.P. Yu, Y. Wu, G. Li, Polymer solar cells with enhanced open-circuit voltage and efficiency. Nat. Photonics **3**, 649–653 (2009)
35. L.J. Huo, J.H. Hou, H.Y. Chen, S.Q. Zhang, Y. Jiang, T.L. Chen, Y. Yang, Bandgap and molecular level control of the low-bandgap polymers based on 3,6-dithiophen-2-yl-2,5-dihydropyrrolo[3,4-c]pyrrole-1,4-dione toward highly efficient polymer solar cells. Macromolecules **42**, 6564–6571 (2009)
36. R.C. Coffin, J. Peet, J. Rogers, G.C. Bazan, Streamlined microwave-assisted preparation of narrow-bandgap conjugated polymers for high-performance bulk heterojunction solar cells. Nat. Chem. **1**, 657–661 (2009)

37. X. Gong, M. Tong, Y. Xia, W. Cai, J.S. Moon, Y. Cao, G. Yu, C.-L. Shieh, B. Nilsson, A.J. Heeger, High-detectivity polymer photodetectors with spectral response from 300 nm to 1450 nm. Science **325**, 1665–1667 (2009)
38. J.Y. Kim, K. Lee, N.E. Coates, D. Moses, T.Q. Nguyen, M. Dante, A.J. Heeger, Efficient tandem polymer solar cells fabricated by all-solution processing. Science **317**, 222–225 (2007)
39. S. Sista, M.-H. Park, Z. Hong, Y. Wu, J. Hou, W.L. Kwan, G. Li, Y. Yang, Highly efficient tandem polymer photovoltaic cells. Adv. Mater. **22**, 380–383 (2010)
40. Y.Y. Liang, Y. Wu, D.Q. Feng, S.T. Tsai, H.J. Son, G. Li, L.P. Yu, Development of new semiconducting polymers for high performance solar cells. J. Am. Chem. Soc. **131**, 56 (2009)
41. Proprietary cylindrical modules by Solyndra, http://www.solyndra.com/technology-products/cylindrical-module
42. M.A. Green, K. Emery, D.L. King, S. Igari, W. Warta, Solar cell efficiency tables – (version 26). Prog. Photovoltaics **13**, 387–392 (2005)
43. C. Piliego, T.W. Holcombe, J.D. Douglas, C.H. Woo, P.M. Beaujuge, J.M.J. Freĭchet, Synthetic control of structural order in N-Alkylthieno[3,4-c]pyrrole-4,6-dione-based polymers for efficient solar cells. J. Am. Chem. Soc. **132**, 7595–7597 (2010)
44. Y. Zhang, S.K. Hau, H.L. Yip, Y. Sun, O. Acton, A.K.Y. Jen, Efficient polymer solar cells based on the copolymers of benzodithiophene and thienopyrroledione. Chem. Mater. **22**, 2696–2698 (2010)
45. Y.P. Zou, A. Najari, P. Berrouard, S. Beaupre, B.R. Aich, Y. Tao, M. Leclerc, A thieno[3,4-c]pyrrole-4,6-dione-based copolymer for efficient solar cells. J. Am. Chem. Soc. **132**, 5330 (2010)

Chapter 6
Integrated Optical Confinement Geometry Device

In 2009, our group first looked for a large area fiber cell by bundling several fibers together (in Fig. 6.1). All fibers surrounded by Al cathode and ITO anodes at the top of fibers are connected to each other by silver paste. This idea inspired us to design more and better integrated OCGPVs, including a stamped fiber matrix on planar plastic substrate, domes or cones substrate and even ZnO nano-rod by near field approach, which will be introduced in this chapter.

6.1 Fiber Bundles

Of course, optical fiber processing is a very advanced science and a first thought was to simply draw-coat the fibers using existing technology, then use post coating to cut and align fiber segments. This presents very challenging problems of contacting the inner electrode as well as having control over the fiber alignment. Early devices made from 8 to 10 fiber-cells were arranged in parallel as a bundle and using silver paste contacts did work, but was so time intensive that this concept was dropped early. Nevertheless, they did show that the fiber-device performance was additive and, the angular response and high capture efficiency of the individual fibers were reflected in the bundles as well. Finally, the effects of the packing fraction between fibers lowered performance by approximately ¼.

6.2 Aligned Stamped Fiber Devices

Obviously, the fabrication of large area devices in the fiber-device configuration is the greatest challenge to the use of such technologies. One approach to large area devices has recently been demonstrated that utilizes aligned fiber arrays stamped into plastic substrates in a fashion similar to that of a hair brush. The substrates can then be coated in a variety of ways: sputter deposition of inner contacts, dip coating

Y. Li, *Three Dimensional Solar Cells Based on Optical Confinement Geometries*, Springer Theses, DOI 10.1007/978-1-4614-5699-5_6,

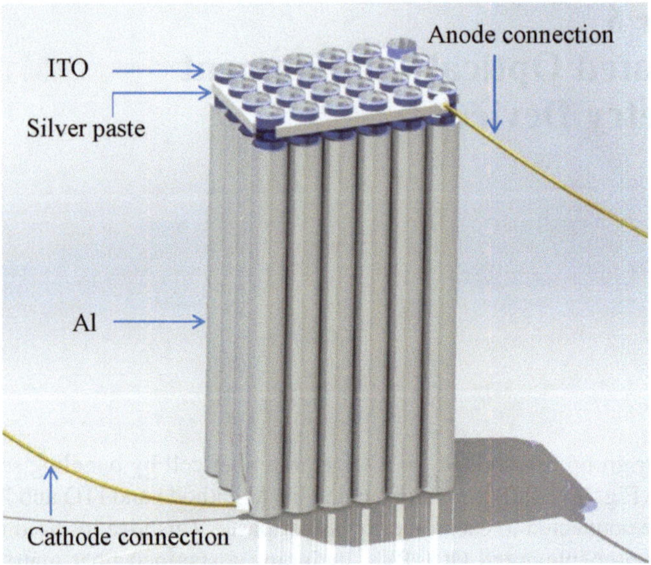

Fig. 6.1 Structures of Fiber cell bundle in which each single fiber cell was connected with each other in parallel by silver paste (ITO-ITO, Al-Al). *Blue* is ITO that is connected by silver paste

or spray coating of active materials, and evaporation of outer contacts. This results in the individual fiber-devices in parallel contact, producing an additive current across the surface as needed. Shown in Fig. 6.2 are two examples of such fiber substrates, one using very short fibers (almost lenses with a $l/d < 1$) and one with longer fibers ($l/d \gg 1$); where l = fiber length and d = fiber diameter.

Fabrication of functioning organic devices on such 3-D substrates can be challenging since typically, tolerances for film thicknesses are rather tight. However, drop casting and spray techniques do allow for simple devices to be made that exhibit reasonable performance. Indeed, the inner conductor places the strictest constraints on performance, not the photoactive absorbers (to which the device seems to be extremely fault tolerant).

6.2.1 *Aligned Stamped Fiber (Dome) Devices*

We reported a new version of fiber-based photovoltaic device built on plastic substrate on which multiple optical fibers were stamped on the surface. Bulk heterojunction organic polymer was spray-deposited between ITO and aluminum all around fibers on the substrate with regioregular poly(3-hexylthiophene) (P3HT) and phenyl-C61-butyric acid methyl ester (PCBM) as absorbing layer. Internal conversion efficiency was tested under standard solar simulation and was found to vary with changing incident angles; 15° incident gave the highest performance

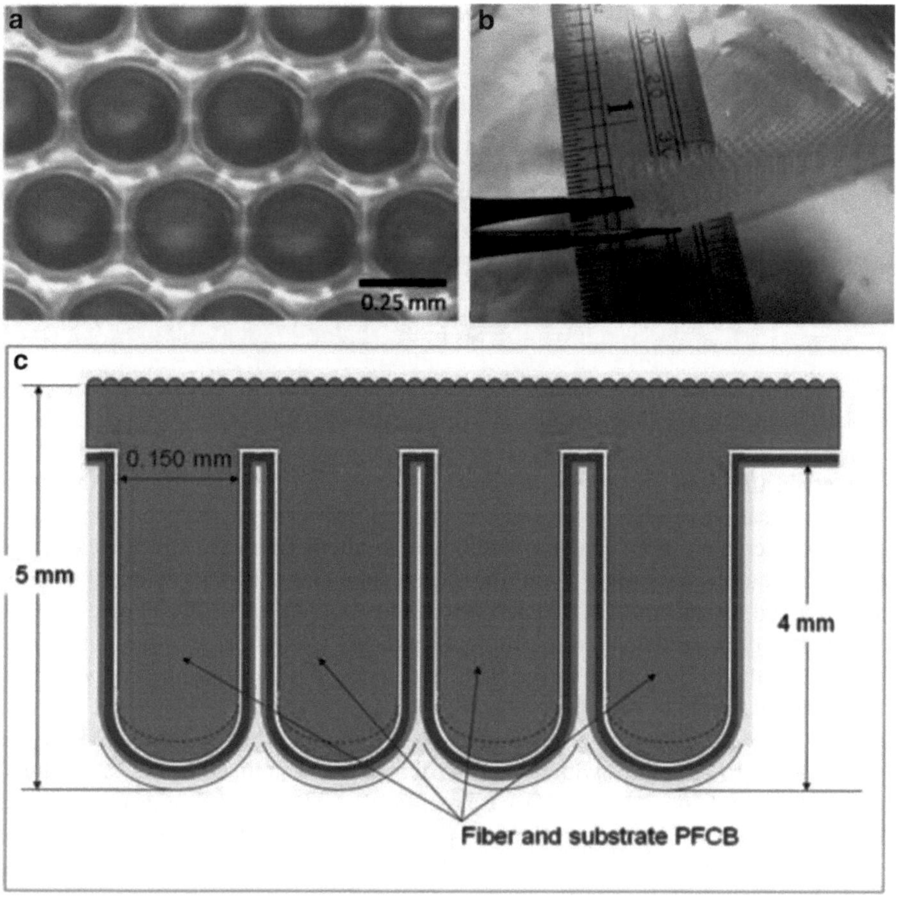

Fig. 6.2 (**a**) Optical microscope image of plastic substrate with stamped fiber on the surface. (**b**) Photo of aligned plastic fibers on fixed substrate. (**c**) A schematic of the cross-section of the device after coating. Light enters from the top

(1.82%) while it was 1.43% when the light came through perpendicularly. The behavior of open circuit voltage (Voc) and fill factor (FF) versus incident angles were examined under the comparison between flat substrate and fiber substrate.

Organic solar cells have been developed with an inexpensive approach to transforming the solar energy into electricity. Recently, the most widely used polymer for the solar absorbing layer was the blend of P3HT:PCBM organic materials, which shows 5% as the highest (Reprinted with permission from [1] Copyright 2010, American Institute of Physics) conversion efficiency from light to current. However, this number hasn't satisfied the requirement of industrial production (10%). Some research focused on low bandgap polymers, with which the internal conversion efficiency was approaching 8% [2]. While others were developing the new electronic structures like the tandem cell [3] or adding dyes to

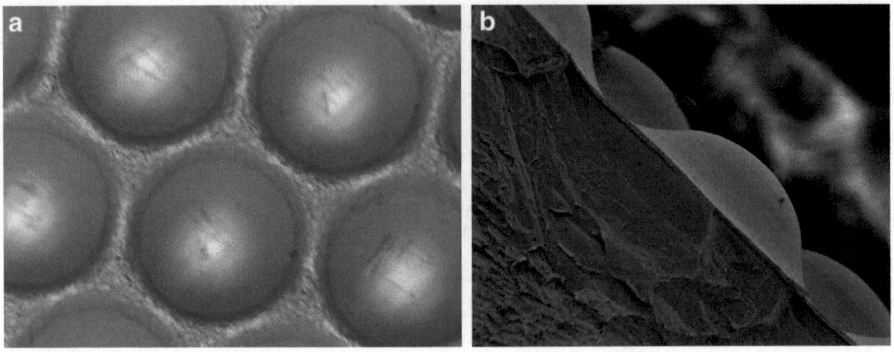

Fig. 6.3 Stamped fiber substrate under (**a**) optical microscope (10×) (**b**) SEM; the size of the fiber was 100 μ (diameter) × 50 μ (length)

strengthen the light absorption in sensitized solar cell (DSSC) [4]. On the other hand, optical structures were also examined. Nowadays fiber based organic photovoltaics have been drawn attention as an alternative structure for capturing the light by a physical way [5]. In fiber based solar cell (FBSC), organic thin films are wrapped around the ITO coated optical fibers concentrically. People test the fiber cell in different ways, for example, Brendan O'Connor, et al. illuminated from one side of the FBSC and achieved 0.5% efficiency [6]; End-illumination was used to couple solar light to the device as in an optical wave guide [7]. However, all those devices were built on single glass fibers with very tiny active areas. We report completely new structures with numerous optical fibers stamped on one substrate to extend this single device into a larger area. Further, all the other approaches could be applied on these substrates like new polymers, tandem, DSSC, etc., to increase the power conversion efficiency. Also, these new substrates were made of plastic (polystyrene), which is a much cheaper and lighter material compared to glass and the flexible solar cell can be realized simply by reducing the thickness of the sub-layer of the substrate Fig. 6.3.

6.2.1.1 Device Fabrication

Polystyrene substrates with and without stamped fibers were cleaned in an ultra-sonically bath with DI water for 30 min and rinsed in isopropyl alcohol flow. After drying in a vacuum oven at 100°C, they were transferred into a vacuum chamber of a (BOSCH) sputtering system and pumped down to 1.7×10^{-5} torr. A 100 nm transparent and conductive indium tin oxide (ITO) film was sputtering deposited in argon and oxygen flow from an ITO target while the substrates were heated up to 80°C.

Due to the 3-D feature of the fiber substrate, spray-coating was chosen as the optimal way of depositing the organic films over the fibers uniformly. After exposing the obtained ITO coated substrate in the ozone plasma for 10 min, a formula of PEDOT:PSS (PH 750 purchased from H. C. Starck) was spray-coated on the

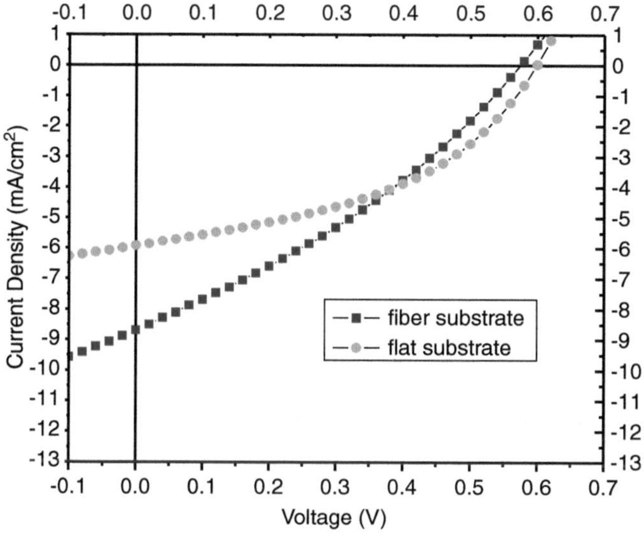

Fig. 6.4 Comparison of IV curve under illumination of stamped short fiber (dome) substrate

surface by a hand-held airbrush (Iwata HP-CB connected with an air compressor). The PEDOT layer was dried for 10 min at 100°C on a hot plate under atmosphere. A diluted blend [8] solution of P3HT:PCBM (1:0.8) in chlorobenzene (1.5 mg/ml) was deposited by airbrush after drying the PEDOT layer; the distance between device and nozzle was 5 cm to keep the sub-layer from redissolving [9]. Finally, the whole device was kept in a vacuum chamber at 1×10^{-6} torr for the thermally evaporated LiF and aluminum with a thickness of 0.3–0.4 nm and 100 nm respectively, as the back contact of the solar cell.

6.2.1.2 Performance

After thermally annealing in a nitrogen filled glove box at 100°C for 10 min, the devices were tested under AM1.5 g standard (Oriel) operating with an illumination intensity of 100 mW/cm^2. The highest internal efficiency was 1.89% while the planar device only achieved 1.51% on the planar substrate made of polystyrene Fig. 6.4.

Solar cells built on the stamped fiber substrates produced higher current density than planar plastic devices when they were built under the same conditions. In the case of the planar device, light came through the whole device vertically and the active layer could only absorb the light twice. But for the FBSC, the round-topped fiber-end would redirect the reflected light to the other part of the absorbing layer, which fully utilized the light. The multiple internal reflections within the fibers strengthened the light capture so that the P3HT:PCBM layer around the fibers was able to absorb incoming light from all directions. The difficulty of covering the fluctuant surface yielded a lower fill factor in Fig. 6.4 of FBSC device which limited the overall efficiency.

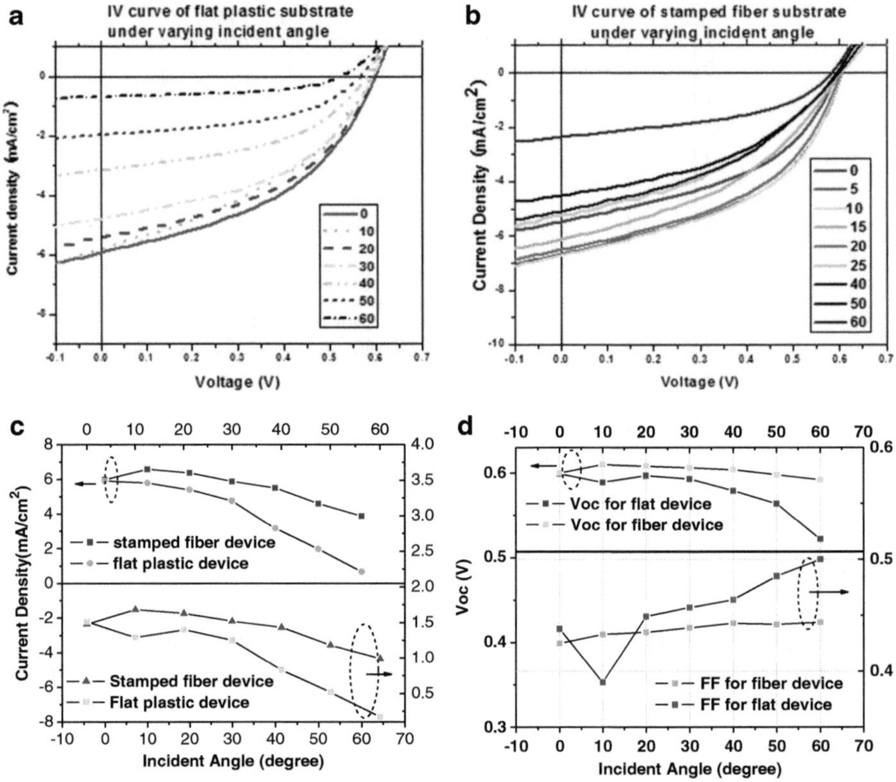

Fig. 6.5 Comparisons of incident angle dependence

By changing the incident angles, we got different solar cell performance for stamped-fiber devices. In Fig. 6.5a, the current density was increasing by the rising angle from 5.46 mA/cm^2 at 0° to 6.72 mA/cm^2 at 10° and dropped to 6.50 mA/cm^2 at 20°. However, both the Voc and fill factor remained stable in Fig. 6.5c for the stamped fiber device, while the flat ones had a rising FF and a decreasing Voc. According to the literature [10], the fill factor would increase when the light density got extremely weak, which explained the larger fill factor at higher angles for the flat device. But the fill factors we got for fiber substrate kept stable when the angle changed. One possible explanation for this kind of tendency is the capability of light-trapping of fibers on the substrate, which means the light density didn't change within the fiber structures.

6.2.2 Aligned Stamped Fiber Devices (Long)

Longer fibers substrate was utilized to improve light confinement by its longer optical path. The stamping process above, as well as numerous photolithographic processes, can be used to give large area aligned arrays of fibers from 0.2 to 2 mm in diameter, with lengths of 0.25–5 mm and separations of no more than 100 μm. We have investigated substrates like these fabricated from poly (methylmethacrylate) (PMMA), and a number of photolithographic resins, all with excellent transmission properties. Since the adjacent fibers block each other, it is not suitable to sputter ITO to surround the fiber; conductive PEDOT:PSS was therefore used as the anode. Because these devices are much longer, internal resistance is very high when ITO is not utilized. Nevertheless, we have begun optimization of such substrates for applications that require both a flexible and rigid solar cell. The upper support of the fiber array is made very thin to achieve flexible device configurations. Thicker upper supports allow for ease in coatings, though also make the device more rigid. Again, using spray coating of P3HT:PCBM, and dip-coated PEDOT as the internal conductor, devices with longer fibers have been fabricated. IV curves from flexible substrates are shown in Fig. 6.6 in which annealed and unannealed devices fabricated with chlorobenzene and toluene are compared.

We note that for these devices, the longer fibers do not allow for efficient use of the lower conductivity PEDOT inner electrode, and consequently, space charge begins to build as evident in Fig. 6.6b as the "hump" in the IV near zero current. Further, as can be seen, device performance is strongly dependent on processing of the absorber in this case. This is not true in the shorter fiber case and probably derives because the longer fibers make the device particularly sensitive to growing internal resistance.

6.3 Cones-Based Fiber Device

Taking all factors above into consideration, we designed a substrate consisting of cones having a fiber-tip, as shown in Fig. 6.7. This architecture has the following advantages:

1. Easy deposit of thin films on substrate, because all the surface is facing upwards, i.e., no dead corner.
2. Focus light on the fiber-tip, to further confine and absorb light.
3. The fiber-tip with nonlinear optical material can make a down-conversion or up-conversion for incident light, generating more current.
4. The area of this architecture is not too large, compared with that of the fiber-bundles or aligned fibers, such that the voltage loss is reduced.

The general form for this architecture is written as below

$$(z \geq b)||\left\{\left[\left(\frac{(x-d_1/2)}{2}\right)^2+\left(\frac{(y-d_2/2)}{2}\right)^2 \leq \left(\frac{z-c}{a}\right)^2\right]\&\&z \geq c\right\}$$
$$||\left[(x-d_1/2)^2+(y-d_2/2)^2 \leq r^2\&\&z \geq (d_3-h)\right] \tag{6.1}$$

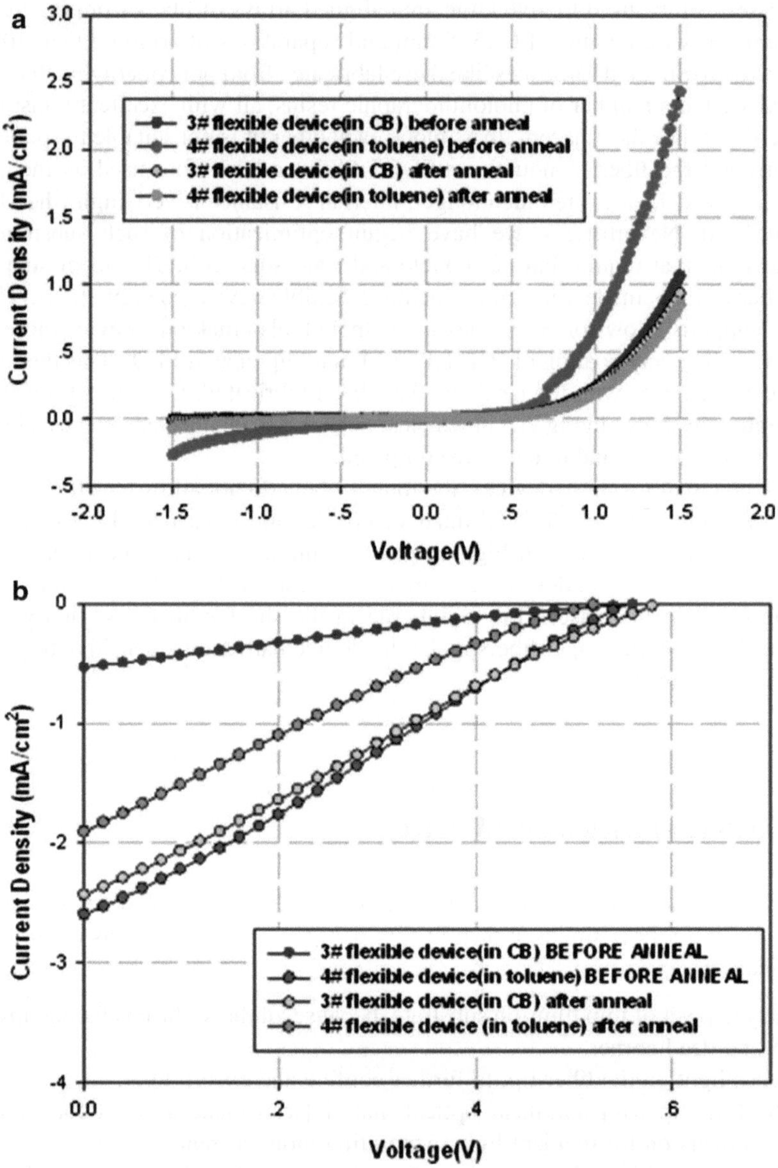

Fig. 6.6 Comparison of FBSC IV curve (**a**) in dark (**b**) under illumination for flexible substrates using different solvents

It is easy to derive other parameters based on the four basic variables a, b, d, h.

$$g = \frac{\sqrt{2(b-c)^2}}{a} \tag{6.2}$$

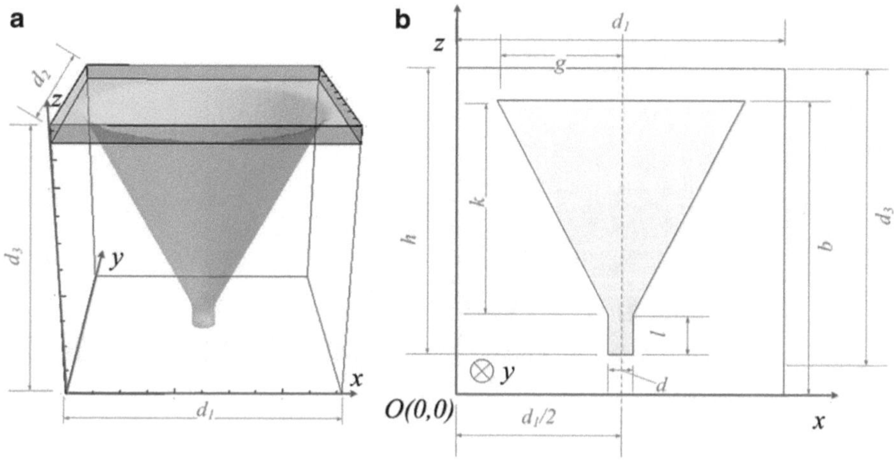

Fig. 6.7 (**a**) 3D drawing of cones-based fiber device (**b**) Sizes in details for this architecture

$$k = [h - (d_3 - b)]\left(1 - \frac{d}{2g}\right) \tag{6.3}$$

$$l = h - k \tag{6.4}$$

Using the simulation program based on the models in Chaps. 1, 2 and 4, we search for the optimum architecture from more than 10,000 combinations of the four variables above. Figure 6.8 shows two chosen architectures and their corresponding sizes for macro scale and micro scale, respectively. The micro-cone doesn't include a fiber on its top because it is too difficult to deposit extra material on this extremely tiny fiber.

Figure 6.9 shows the simulation result of incident angle' dependence to absorption, which is similar to that of planar cells but unlike that of common fiber-based solar cells [5, 7, 11, 12]. Because this architecture is closed at its top, unlike fiber having an open end where light may pass through fiber, light with small incident angle can also be highly absorbed by this structure.

6.4 ZnO Nano-rod Inverted Solar Cell

In this section, a near field approach, ZnO nano-rod (micro-fiber) is introduced to improve the current harvest in planar cells [13]. The thickness of MoO_3 is used to control the light distribution focusing on the active layer, to reach the optimum absorption.

Organic solar cells based on vertically-aligned zinc oxide nanorod arrays (ZNR) in an inverted structure of indium tin oxide(ITO)/ZNR/poly(3-hexylthiophene):

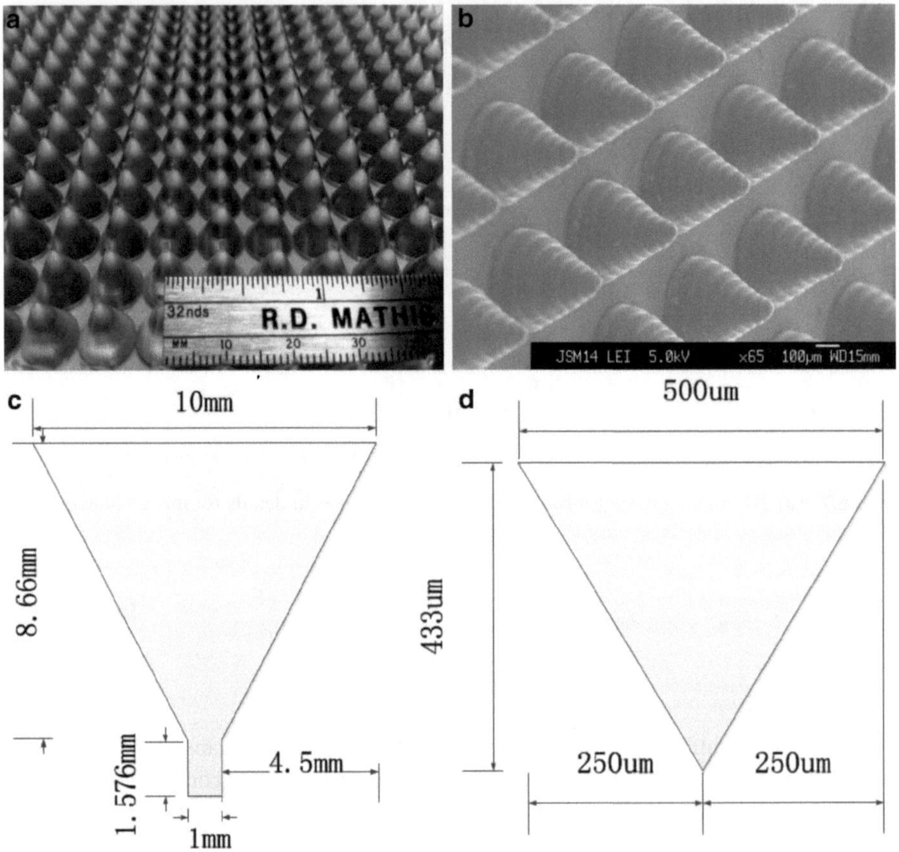

Fig. 6.8 (**a, b**) Are two optimum architectures of macro-scale (centimeter) and micro-scale (micrometer). Their engineering drawings are shown in (**c**) and (**d**) respectively

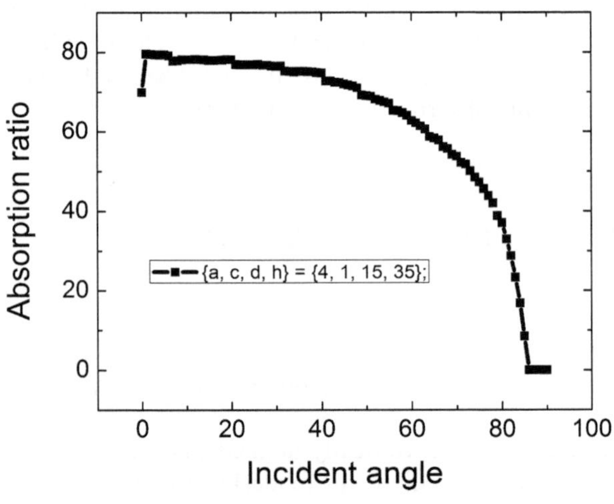

Fig. 6.9 Absorption ratio versus incident angle (simulation)

(6,6)-phenyl C61 butyric acid methyl ester(P3HT:PCBM)/MoO_3/Aluminum(Al) were studied. We found that at the optimum MoO_3 layer thickness condition of 20 nm, the MoO_3 can effectively decrease the probability of bimolecular recombination either at the Al interface or within the active layer itself. For this optimum condition, we get a power conversion efficiency of 2.15%, a short-circuit current density of 9.02 mA/cm^2, an open-circuit voltage of 0.55 V, and a fill factor of 0.44 under 100 mW/cm^2 irradiation. Our investigations also show that the highly crystallized ZNR can create short and continuous pathways for electron transport and increase the contact area between the ZNR and the organic materials.

Photovoltaic (PV) devices based on conjugated polymers such as poly(3-hexylthiophene) (P3HT) and (6,6)-phenyl C61 butyric acid methyl ester (PCBM) have attracted a lot of attention in recent years because they are easily fabricated. This offers the possibility of solar cells that are cost-effective, flexible, and of large area. However, the strong acidic nature of the poly(3,4-ethylene dioxythiophene):(polystyrene sulfonic acid) (PEDOT:PSS) modified layer can lead to corrosion of the indium tin oxide (ITO) electrode [14] and degradation of the conjugated polymer in the active layer. Several groups have constructed inverted structure PV devices with highly transparent metal oxides, i.e., TiO_x [15, 16] or ZnO [17, 18] on the transparent electrode as the cathode buffer layer. ZnO is a good candidate for this application because it is a cheap and environmentally-friendly material that can be synthesized with high purity and crystallinity at low temperature. The high electron mobility and high transparency in the visible-wavelength range allow it to be both an effective electron transporter and excellent wave guide. Olson et al. [19, 20] have reported efficient ZnO/polymer devices with different ZnO morphologies. Vertically oriented ZnO nanorods arrays (ZNR) can provide continuous electron transport pathways. Moreover, the high surface-to-volume of the vertically-aligned ZNR can ensure three-dimensional contact with the polymer and provide many direct electrical pathways to transport electrons to the cathode directly. Vishal Shrotriya et al. [21] have verified that MoO_3 can replace PEDOT:PSS as the hole selective layer, which can be conveniently deposited by evaporation before depositing the anode metal. The thickness of the MoO_3 hole selective layer is a vital parameter in the inverted solar cell structure. To further understand the effect of MoO_3 thickness on the ZNR-based invert structures, layers of various thicknesses were deposited onto a series of inverted ZNR solar cells by thermal evaporation. We found the thickness of the MoO_3 layers had a significant effect on the performance of the inverted ZNR solar cells.

According to the literature [22, 23], the expected energy level diagram of the ZNR based BHJ solar cell is shown in Fig. 6.10. Since the pairs of hole and electron are generated at the interface of donor and acceptor, the hole will travel through the MoO_3 layer while electrons will inject to the conducting band of the ZnO and will be collected by aluminum and ITO respectively as shown in Fig. 6.11.

The ZNR based inverted solar cell was fabricated by the normative process as illustrated in Fig. 6.10. Devices were fabricated on pre-cleaned and etched conducting ITO (Delta Technologies $R_s = 10\ \Omega\ sq^{-1}$) coated glass substrates. A 130 nm P3HT:PCBM (1:0.8 Weight ratio, 15 mg/ml in chlorobenzene) layer

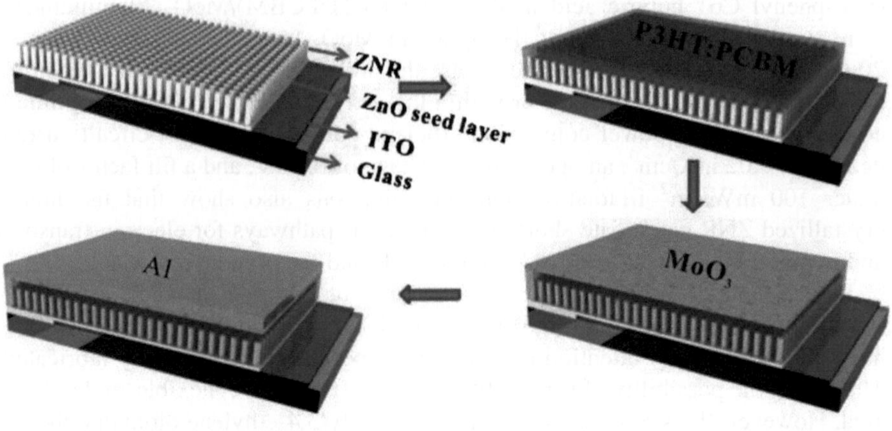

Fig. 6.10 Schematic illustration of the fabrication process of ZNR based BHJ solar cells

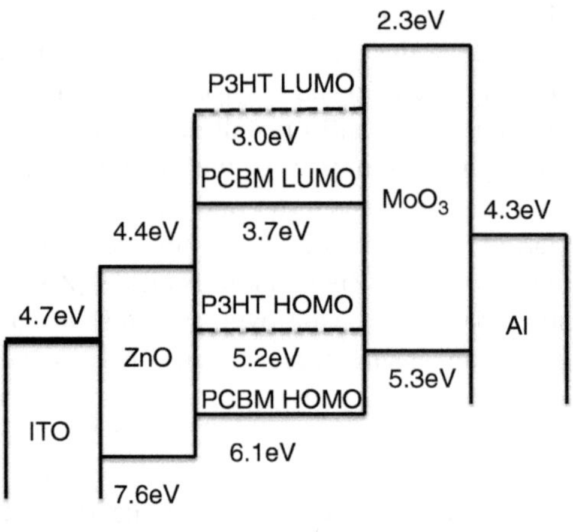

Fig. 6.11 The estimated energy diagram of the inverted cell

was spin-cast on top of pre-grown ZNR [24] substrates. Different thicknesses of MoO_3 were deposited and capped with a 100 nm Al by thermal evaporation (5×10^{-6} torr) with a shadow mask. The devices were then annealed under N_2 ambient at 150°C for 7 min. The photovoltaic characteristics were measured with a Keithley 236 source-measurement unit under 100 mW/cm^2 using an AM 1.5G standard Newport #96000 Solar Simulator. The external quantum efficiency (EQE) was measured using a Newport Cornerstone 260 Monochromator in conjunction with a Newport 300W Xenon light source, and a Merlin Lock-In amplifier. The morphology analysis of the ZNRs and solar cell composite structure were each done with a Scanning Electron Microscope (JEOL, JSM-6330 F).

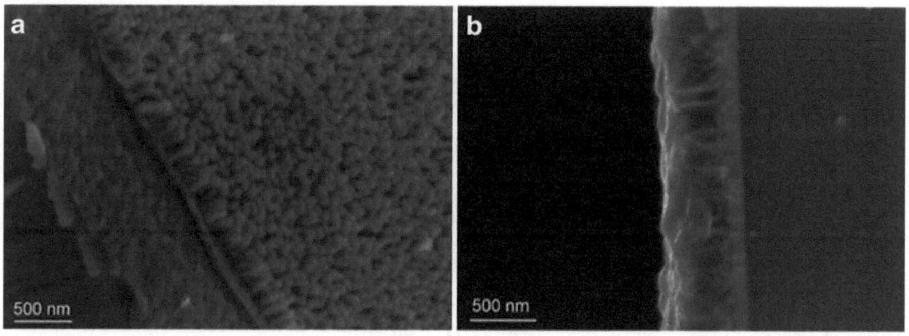

Fig. 6.12 (**a**) The side view SEM images of the ZNRs grown by the hydrothermal method, (**b**) cross sectional views of the P3HT: PCBM-coated ZNRs

Figure 6.12a shows ZNRs approximately 120 nm in length that are grown on a 100 nm ZnO seed layer. The ZNRs form a highly organized structure with the rods forming perpendicular to the seed layer in closely packed formation, leading to a rough surface. From Fig. 6.12b we can see that P3HT: PCBM formed an even coating on top of the ZNRs, and infiltrated well between the rods.

Figure 6.13a shows the J-V curve of devices without (0 nm: device A) and with 5, 10, 15, 20, and 25 nm MoO_3 blocking layer (device B, C, D, E, and F respectively), we can see that with the increased thickness of the MoO_3 layer, the short circuit current density (J_{sc}) increased from 5.25 to 9.02 mA/cm^2. It should be noted that the surface roughness of the ZNR array is much larger than a spin coated sol gel based ZnO film. This led us to use very slow spinning speeds to ensure a reasonably even coating of polymer-fullerene solution on the rods. However, the surface of the active layer was still not smooth enough for a 15 nm or less MoO_3 layer to form a consistent film. For device B and C, the MoO_3 layer could only cover the top of each rod and the bottom well between two ZnO nanorods, leaving the aluminum capping layer in direct contact with the polymer at the sides of the rods. This means that the MoO_3 is only partially effective as a hole transport layer. Due to the similar conduction band edge of ZnO and Al, the electron can jump from the lowest unoccupied molecular orbital of PCBM to either the conduction band of the Al or the ZnO. This decrease in electrode selectivity can effectively increase the probability of bimolecular recombination (either at the aluminum interface or within the active layer itself). For thicknesses between 15 and 20 nm of MoO_3 more and more of the surface of these rods were covered until there was no direct connection between the active layer and the cathode. The MoO_3 could then act as a full-time hole transporting, electron and exciton blocking layer, which led to a significant increase in J_{sc}. The open-circuit voltage (V_{oc}) of the 20 nm MoO_3 device is increased from 0.223 to 0.555 V, providing further evidence for removal of shunts from the active layer to the aluminum cathode. For device F, a thicker MoO_3 layer reduced the shunt resistance and caused the mismatch of optical field distribution in active layer [25, 26], which resulted in a decrease in the performance of the device.

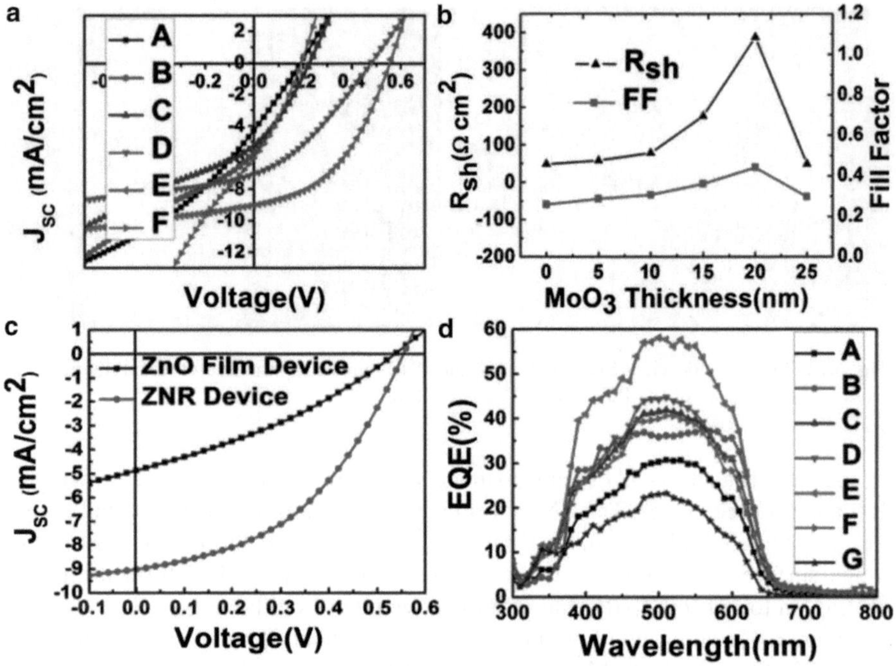

Fig. 6.13 (**a**) J-V curve of the ZNR based inverted solar cell with different MoO_3 thickness and (**b**) the dependence of FF on a R_{sh} (**c**) *J-V* curve of ZNR and ZnO film based inverted solar with 20 nm MoO_3 layer (**d**) The EQE of both the ZNR with different MoO_3 thickness (ZNR devices) and ZnO film with 20 nm MoO_3 layer devices (ZnO device)

The extracted device parameters are summarized in Table 6.1. We can see that the fill factor (FF) of the device increased with the increase of MoO_3 thickness. However, the FF is still very low compared with that reported [27], because we get relatively low shunt resistance(R_{sh}) due to the presence of electrical shunt paths created by the ZnO nano rod arrays and the voltage dependent carrier recombination created by the rough surface of the ZNR. An extreme drop in R_{sh} can cause a drop in the V_{oc} of a solar cell. The fill factor of solar cells can be influenced by the non-zero R_s and by the finite shunt resistance (R_{sh}). According to literature [28, 29], R_{sh} and R_s can be independently derived from the slope of the I-V curve of a solar cell at $I = I_{sc}$ ($V = 0$) and $V = V_{oc}$ ($I = 0$). In this case, as we can see in Table 6.1, all the devices have an approximate R_s around 30 Ω/cm^2, so we propose that the main influence on the performance of these cells is the R_{sh}. which can be seen in the Fig. 6.13b. Increasing the R_{sh} results in an increase in FF. We should note that even though, in our device, the electron transport layer contains both a 100 nm ZnO seed layer and a 120 nm nanorod array layer, we still can get a higher J_{sc} than the flat ZnO film device reported in literature [27], that may ascribe to the highly ordered ZNR providing a more direct electron pathway.

Table 6.1 The summary of performance of ZnO nano-rod inverted solar cells

	EF(%)	V_{oc}(V)	J_{sc}(mA/cm^2)	FF	EQE (%)	R_{sh}(Ω cm^2)	R_s(Ω cm^2)
A	0.203	0.188	4.29	0.260	21.7	48.08	30.35
B	0.394	0.223	6.12	0.288	36.4	58.28	30.26
C	0.367	0.277	5.43	0.307	41.1	78.11	30.29
D	1.06	0.478	6.99	0.326	43.4	176.99	46.87
E	2.15	0.555	9.02	0.442	57.6	387.88	22.17
F	0.343	0.191	6.03	0.298	40.6	49.02	22.44

A device with a 200 nm ZnO seed layer as the electron selective layer was fabricated, for comparison. The other conditions are the same as the 20 nm MoO_3 ZNR device. The J-V curve is shown in Fig. 6.13c. We can see that the V_{oc} of the two devices are similar, but the J_{sc} with only the ZnO film is much lower than the ZNR device. It is reported that the diffusion length of excitons in the conjugated polymer is typically on the order of ~10 nm [30], so the photo generated charge carriers in the 130 nm polymer layer in the ZnO film device are mostly recombined before collection and transport to the electrode in the polymer. Because the gap between two rods are on the order of 10 nm, which is suitable for efficient charge transport, each ZnO nanorod can create short and continuous pathways for electron transport and increase the contact area between the ZnO nanorod arrays and the organic materials, resulting in a higher photocurrent. Our experimental results confirm this explanation.

The external quantum efficiency (EQE) of both the ZNR with different MoO_3 thickness (ZNR devices) and ZnO film with 20 nm MoO_3 layer devices (ZnO device) are shown Fig. 6.13d. The ZNR devices show a maximum EQE of 57.6% at 515 nm. The peak in the EQE spectrum at about 350 nm results from the absorption of the ZnO. With the increased thickness of the MoO_3 layer, a dramatic increase in the EQE is observed of around 36.4%. This increase is likely the result of enhanced charge collection in the device. All of the ZNR devices have a higher EQE than the ZnO device. The EQE spectrum is broader for the ZNR devices in comparison to the ZnO device.

To further confirm our experimental results, we also use a transfer matrix method to calculate the optical field distribution in the active layer with different thickness of MoO3 layer (A detailed simulation is based on Sect. 1.2.2 for optical field distribution). The results are shown in Fig. 6.14, and we can see that the highest absorption peak is around 490 nm when MoO_3 layer is 32 nm. The little difference with the experimental results is due to the fact that our simulation is based on the optical field, due to the semi-insulator property of the MoO_3 material, and the resistance is too hard to be measured by the Four-point Probe Method for a 20 nm film. In our device, a thicker MoO_3 layer will greatly increase the R_s.

Thus, we can deduce that the thickness of MoO_3 plays a key role in a ZNR based inverted solar cell, and the ZNR based solar cell is a promising candidate for a high efficiency PV device. To further improve the performance of the ZNR based PV device, we should improve the surface roughness of the ZNRs to get a high FF, which has great potential to improve the efficiency of a ZNR based solar cell.

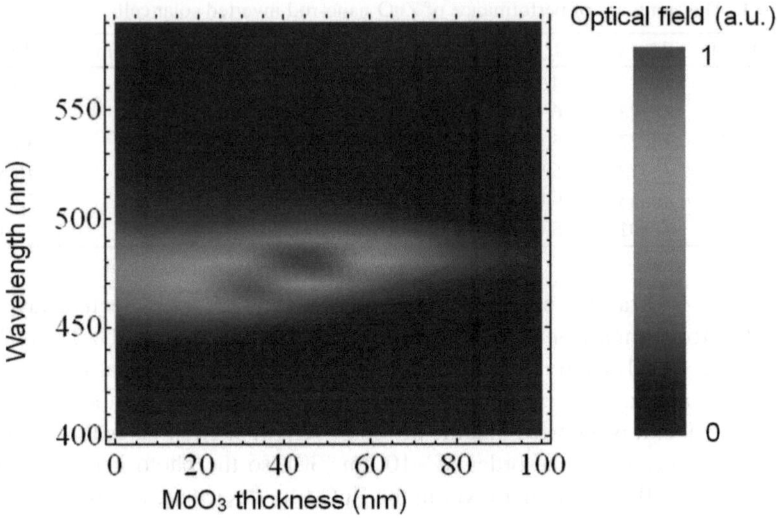

Fig. 6.14 Calculated optical field for different thickness of MoO_3 layer

In summary, we studied the effect of the MoO_3 thickness on a ITO/ZNR/P3HT: PCBM/MoO_3/Al inverted solar cell. We got a maximum efficiency of 2.15% with a 20 nm MoO_3 layer, which can effectively improve the interface between the active layer and the Al electrode. Our investigations also show that the vertically oriented, high electron mobility ZNR arrays can provide continuous electron transport pathways and increase the contact area between the ZNR and the organic materials, making the ZNR based solar cell a promising candidate for high efficiency PV devices.

References

1. S. Gunes, H. Neugebauer, N.S. Sariciftci, Conjugated polymer-based organic solar cells. Chem. Rev. **107**, 1324–1338 (2007)
2. L.P. Yu, Y.Y. Liang, Z. Xu, J.B. Xia, S.T. Tsai, Y. Wu, G. Li, C. Ray, For the bright future-bulk heterojunction polymer solar cells with power conversion efficiency of 7.4%. Adv. Mater. **22**, E135 (2010)
3. K. Lee, J.Y. Kim, N.E. Coates, D. Moses, T.Q. Nguyen, M. Dante, A.J. Heeger, Efficient tandem polymer solar cells fabricated by all-solution processing. Science **317**, 222–225 (2007)
4. L.D. Wang, B. Li, B.N. Kang, P. Wang, Y. Qiu, Review of recent progress in solid-state dye-sensitized solar cells. Sol. Energ. Mat. Sol. C. **90**, 549–573 (2006)
5. Y. Li, W. Zhou, D. Xue, J. Liu, E.D. Peterson, W. Nie, D.L. Carroll, Origins of performance in fiber-based organic photovoltaics. Appl. Phys. Lett. **95**, 203503 (2009)
6. B. O'Connor, K.P. Pipe, M. Shtein, Fiber based organic photovoltaic devices. Appl. Phys. Lett. **92**, 193306 (2008)
7. J.W. Liu, M.A.G. Namboothiry, D.L. Carroll, Fiber-based architectures for organic photovoltaics. Appl. Phys. Lett. **90**, 063501 (2007)

8. R. Green, A. Morfa, A.J. Ferguson, N. Kopidakis, G. Rumbles, S.E. Shaheen, Performance of bulk heterojunction photovoltaic devices prepared by airbrush spray deposition. Appl. Phys. Lett. **92**, 033301 (2008)
9. D.Y. Kim, D.J. Vak, S.S. Kim, J. Jo, S.H. Oh, S.I. Na, J.W. Kim, Fabrication of organic bulk heterojunction solar cells by a spray deposition method for low-cost power generation. Appl. Phys. Lett. **91**, 81102 (2007)
10. Y. Li, M. Wang, H. Huang, W. Nie, Q. Li, E.D. Peterson, R. Coffin, G. Fang, D.L. Carroll, Influence on open-circuit voltage by optical heterogeneity in three-dimensional organic photovoltaics. Phys. Rev. B. **84**, 085206 (2011)
11. Y. Li, E.D. Peterson, H. Huang, M. Wang, D. Xue, W. Nie, W. Zhou, D.L. Carroll, Tube-based geometries for organic photovoltaics. Appl. Phys. Lett. **96**, 243505 (2010)
12. Y. Li, W. Nie, J. Liu, A. Partridge, D.L. Carroll, The optics of organic photovoltaics: Fiber-based devices. IEEE J. Sel. Top. Quant. Electron. **16**, 1827–1837 (2010)
13. Y. Li, E. D. Peterson, H. Huang, M. Wang, D. Xue, W. Nie, W. Zhou, and D. L. Carroll, Tube-based geometries for organic photovoltaics. Appl. Phys. Lett. **96**, 243505
14. M.P. de Jong, L.J. van IJzendoorn, M.J.A. de Voigt, Stability of the interface between indium-tin-oxide and poly(3,4-ethylenedioxythiophene)/poly(styrenesulfonate) in polymer light-emitting diodes. Appl. Phys. Lett. **77**, 2255–2257 (2000)
15. M.N. Shan, S.S. Wang, Z.Q. Bian, J.P. Liu, Y.L. Zhao, Hybrid inverted organic photovoltaic cells based on nanoporous TiO_2 films and organic small molecules. Sol. Energ. Mat. Sol. C. **93**, 1613–1617 (2009)
16. G.K. Mor, S. Kim, M. Paulose, O.K. Varghese, K. Shankar, J. Basham, C.A. Grimes, Visible to near-infrared light harvesting in TiO_2 nanotube array-P3HT based heterojunction solar cells. Nano Lett. **9**, 4250–4257 (2009)
17. M.S. White, D.C. Olson, S.E. Shaheen, N. Kopidakis, D.S. Ginley, Inverted bulk-heterojunction organic photovoltaic device using a solution-derived ZnO underlayer. Appl. Phys. Lett. **89**, 143517 (2006)
18. J.P. Liu, S.S. Wang, Z.Q. Bian, M.N. Shan, C.H. Huang, Inverted photovoltaic device based on ZnO and organic small molecule heterojunction. Chem. Phys. Lett. **470**, 103–106 (2009)
19. D.C. Olson, Y.J. Lee, M.S. White, N. Kopidakis, S.E. Shaheen, D.S. Ginley, J.A. Voigt, J.W.P. Hsu, Effect of ZnO processing on the photovoltage of ZnO/poly(3-hexylthiophene) solar cells. J. Phys. Chem. C **112**, 9544–9547 (2008)
20. D.C. Olson, S.E. Shaheen, R.T. Collins, D.S. Ginley, The effect of atmosphere and ZnO morphology on the performance of hybrid poly(3-hexylthiophene)/ZnO nanofiber photovoltaic devices. J. Phys. Chem. C **111**, 16670–16678 (2007)
21. V. Shrotriya, G. Li, Y. Yao, C.W. Chu, Y. Yang, Transition metal oxides as the buffer layer for polymer photovoltaic cells. Appl. Phys. Lett. **88**, 073058 (2006)
22. V. Shrotriya, G. Li, Y. Yao, C.W. Chu, Y. Yang, Transition metal oxides as the buffer layer for polymer photovoltaic cells. Appl. Phys. Lett. **88**, 73508 (2006)
23. W.J.E. Beek, M.M. Wienk, R.A.J. Janssen, Efficient hybrid solar cells from zinc oxide nanoparticles and a conjugated polymer. Adv. Mater. **16**, 1009 (2004)
24. M.J. Wang, G.J. Fang, L.Y. Yuan, H.H. Huang, Z.H. Sun, N.S. Liu, S.H. Xia, X.Z. Zhao, High optical switching speed and flexible electrochromic display based on WO3 nanoparticles with ZnO nanorod arrays' supported electrode. Nanotechnology **20**, 185304 (2009)
25. D.W. Zhao, P. Liu, X.W. Sun, S.T. Tan, L. Ke, A.K.K. Kyaw, An inverted organic solar cell with an ultrathin Ca electron-transporting layer and MoO3 hole-transporting layer. Appl. Phys. Lett. **95**, 153304 (2009)
26. D.W. Zhao, S.T. Tan, L. Ke, P. Liu, A.K.K. Kyaw, X.W. Sun, G.Q. Lo, D.L. Kwong, Optimization of an inverted organic solar cell. Sol. Energ. Mat. Sol. C. **94**, 985–991 (2010)
27. A.K.K. Kyaw, X.W. Sun, C.Y. Jiang, G.Q. Lo, D.W. Zhao, D.L. Kwong, An inverted organic solar cell employing a sol–gel derived ZnO electron selective layer and thermal evaporated MoO3 hole selective layer. Appl. Phys. Lett. **93**, 221107 (2008)

28. K.I. Ishibashi, Y. Kimura, M. Niwano, An extensively valid and stable method for derivation of all parameters of a solar cell from a single current–voltage characteristic. J. Appl. Phys. **103**, 094507 (2008)
29. S.S. Hegedus, W.N. Shafarman, Thin-film solar cells: Device measurements and analysis. Prog. Photovoltaics **12**, 155–176 (2004)
30. K.M. Coakley, M.D. McGehee, Conjugated polymer photovoltaic cells. Chem. Mater. **16**, 4533–4542 (2004)

Chapter 7
Hybrid Optical Confinement Geometry Device

Apart from the more distributional solid state photovoltaic devices, including silicon III-V, polymer, CIGS and CZTS, another type of solar cell based on solutions such as dye-sensitized solar cells has attracted many groups in the past 10 years, and also proved to perform at levels higher than 10% [1]. Some optical confinement geometries with tubes or other cavities connecting with fibers could work well by combining solid-based cells and solution-based materials together, which is known as a hybrid optical confinement geometry solar cell. Furthermore, by means of the feature of applicability working at very high light intensity, a fiber based hybrid solar power system combining photovoltaic and thermoelectricity is designed to reach a very high efficiency over 50%. Yet, this technology is able to be applied to most of photovoltaic materials.

7.1 Photovoltaic-Thermal Solar Energy Collectors Based on Optical Tubes

In this section, we examine the use of oil-filled tubular optical concentrators coupled with a model organic bulk hetrerojunction photovoltaic: poly-3-hexathiophene-[6, 6]-phenyl-C61-butyric-acid methyl-ester(P3HT:PCBM) to create a photovoltaic-photothermal hybrid solar collector. The organic photovoltaic cells were fabricated onto one half of a tubular light pipe and then silicone oil was flowed inside the pipe. This allows solar energy in the visible wavelengths to be effectively converted into electricity by photocell while simultaneously, the silicone oil captures the infrared radiation (IR) part of the spectrum as heat energy. The VIS-IR power conversion efficiency for this model organic system, under normally incident AM1.5 G illumination was found to be: PCE ~ 28%, which is combined by the photovoltaic efficiency (PCE ~ 2%) and the photothermal efficiency (PCE ~ 26%). We further show that the oil filled tube acts as a passive optical element that concentrates

Y. Li, *Three Dimensional Solar Cells Based on Optical Confinement Geometries*, Springer Theses, DOI 10.1007/978-1-4614-5699-5_7,

the light onto the photovoltaic and thereby increases its overall efficiency, but also the range of incident angles in which the light is efficiently captured (Reprinted from [2].Copyright (2011), with permission from Elsevier).

The total number of photons that can be captured by a photovoltaic, and thus the resulting power conversion efficiency (PCE), depends on the threshold photon energy below which electricity conversion does not take place, i.e. the bandgap of the photoabsorber. So, long wavelength photons typically dissipate their energy as heat in photocells instead of generating power. This brings other undesirable consequences to the solar cells as well, such as a drop in the cell efficiency due to the added heat and even permanent structural damage of the cells [3]. To better harness the solar energy, it is necessary to collect the solar energy in the long-wave band of the sun's spectrum. However, trying to use more and more narrow bandgap semiconductors to absorb longer and longer wavelength light results in diminishing voltage output for the solar cell. Thus, the most obvious way to harness this IR solar energy is to transform this part of the spectrum into heat and collect that heat, i.e. photothermal conversion [4]. At present, there are a number of efforts which have demonstrated combined photovoltaic-thermal (PV/T) solar energy collectors, but these have been based primarily on planar geometries and silicon absorbers [3–8].

These first generation hybrid photovoltaic-thermal solar energy collectors (HPVTSC) have resulted in some limited success, but their architectures have not been optimized for either solar or thermal collection of energy. That is, such PV/T devices are typically designed in a planar geometry, yielding relatively low combined photovoltaic/thermal conversion performance. In this work, we examine the use of optical tube concentrators, around which a photovoltaic cell has been added. The resulting hybrid photovoltaic-thermal solar energy collector geometry allows for optimized heating of a thermal fluid added to the tube and for concentration of the light onto the photovoltaic which yields great photovoltaic efficiencies. Our demonstration utilizes a solution-processed bulk heterojunction (BHJ) solar cell fabricated on one half of the tube's circumference and silicone oil contained inside the tubes as the photothermal collector. To fabricate these devices, we used a common absorber system P3HT: PCBM as a model absorber, which has been successfully applied to the fiber-based and tube-based solar cells recently [9–11]. Thus, in these organic-PV/T (OPV/T) devices we utilize the advantages of organic photovoltaic (OPV) materials such as low cost, conformal flexibility, and abundant availability [12]. In order to better harness the utilization of the thermal fluid, we use a relatively smaller tube (1.5 mm I.D.) to get a higher ratio of cross-sectional area to volume. So in this case we build the OPV device directly onto the tube rather than adding a roll-to-roll (R2R) producible flexible OPV with an adhesive [13–16]. However, we find that this architecture further exhibits an increase in the performance of the P3HT:PCBM cell due to the concentrator effect of the tube acting as a focusing lens. Therefore, the combined power collection for this system; which is essentially an organic photovoltaic, is surprisingly high (PCE ~ 28%) compared to around 5% for the P3HT:PCBM system [17]. We suggest that this approach may provide a more rapid approach to commercial viability of organic photovoltaics.

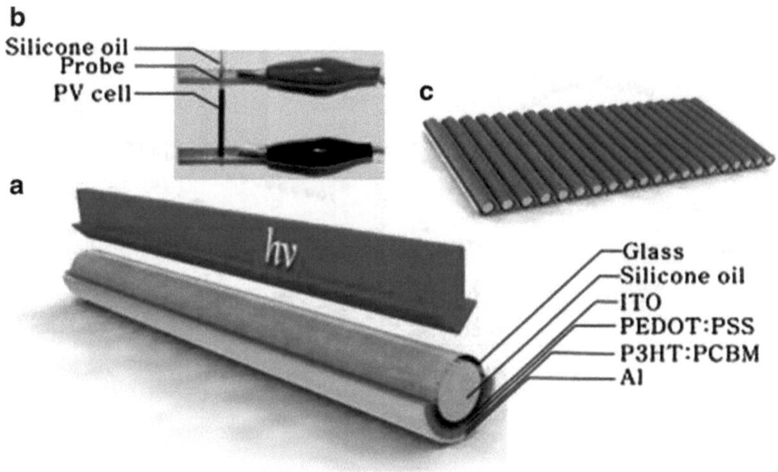

Fig. 7.1 (**a**) Schematic diagram of one tube-based solar energy collector. (**b**) The photographic image of the PV/T solar energy collector. (**c**) Schematic diagram of large area tube solar energy collectors

7.1.1 Fabrication of Tube-Based PV/T Devices

Tube-based PV/T devices were fabricated on glass tubes with one end closed in a hemispherical cap (Chemglass, 1.5 mm I.D., 1.8 mm O.D.). Before building the devices, the tubes were cleaned in an ultrasonic bath. Firstly, the ITO films with a thickness of 100 nm were deposited on the tubes by radio frequency (rf) magnetron sputtering (BOSCH) from an ITO target. Then, these tubes were exposed to ozone for 90 min. Subsequently, by a dip coating procedure, PEDOT: PSS (Clevios p, thickness ~200 nm), P3HT: PCBM (1:0.8 in WT, of 12 mg/ml in chlorobenzene, thickness ~150 nm) were deposited on the tubes. Al electrodes were deposited via thermal evaporation at the pressure of 10^{-6} torr. The length of the tube with active area was 1.8 cm. The resulting device structure is show in Fig. 7.1.

Apertures were used to ensure that side illumination did not occur. The photothermal characteristics of these devices were measured by a K-type thermocouple probe immersed in the inner silicone fluid and heating times of the silicone inner fluid was measured with a stopwatch. The angle of incidence of the light was varied by rotating the tube with respect to the light source. IV and thermal data was collected for a variety of angles.

7.1.2 Performance of Tube-Based PV/T Devices

First Fig. 7.2 examined the effects of the addition of silicone oil into the tube on the angular response of the photocell on the back of the tube. From Fig. 7.2a, we observed that by injecting silicone oil inside the tube, the short current density (Jsc)

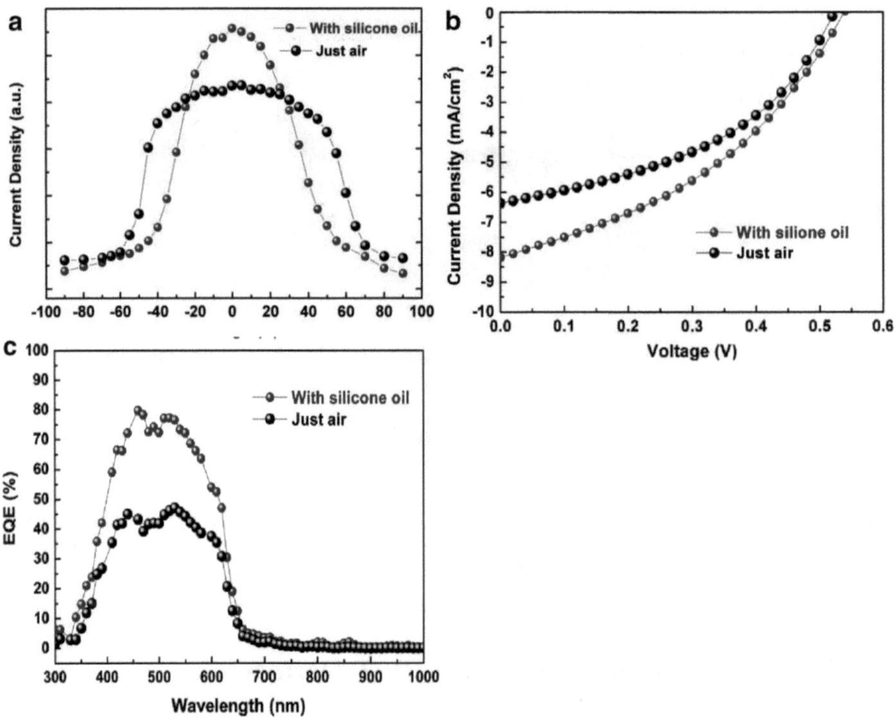

Fig. 7.2 (**a**) The behavior of Jsc versus illumination angles. (**b**) J-V characteristics of the photovoltaic cells under normal (0°) illumination. (**c**) The EQE of the devices with and without silicone oil

of the photovoltaic was generally enhanced over tubes without silicone oil. This enhancement is seen across a 50° angular span, but is most pronounced when the device was under illumination that is normal to the center of the semi-cylinder of the tubular photovoltaic designated as 0° (~30% enhancement), whose current density-voltage (*J-V*) characteristics are shown in Fig. 7.2b. This enhanced performance was confirmed in the EQE measurements taken at 0°, which are shown in Fig. 7.2c. To understand the enhancement of the photovoltaic cell current response after injecting silicone oil, it's necessary to examine the differences in light distribution in the tube with and without silicone oil. To do this, devices were built without the photovoltaic absorber and back contact and they were illuminated from one side. It is apparent visually, from optical images, that the beam from the solar simulator has been strongly focused when the tube is filled with silicone oil, which is shown in the inset of Fig. 7.3b. This is easily understood with the schematic diagram of the light paths of one light ray in the tube with and without silicone oil, which is shown in Fig. 7.3a. So the beam is better focused by injecting higher refraction index materials into the tube, in our case silicone oil.

To further confirm this, we also constructed a mathematical model to calculate the light absorption of the photovoltaic cells with and without silicone oil, where

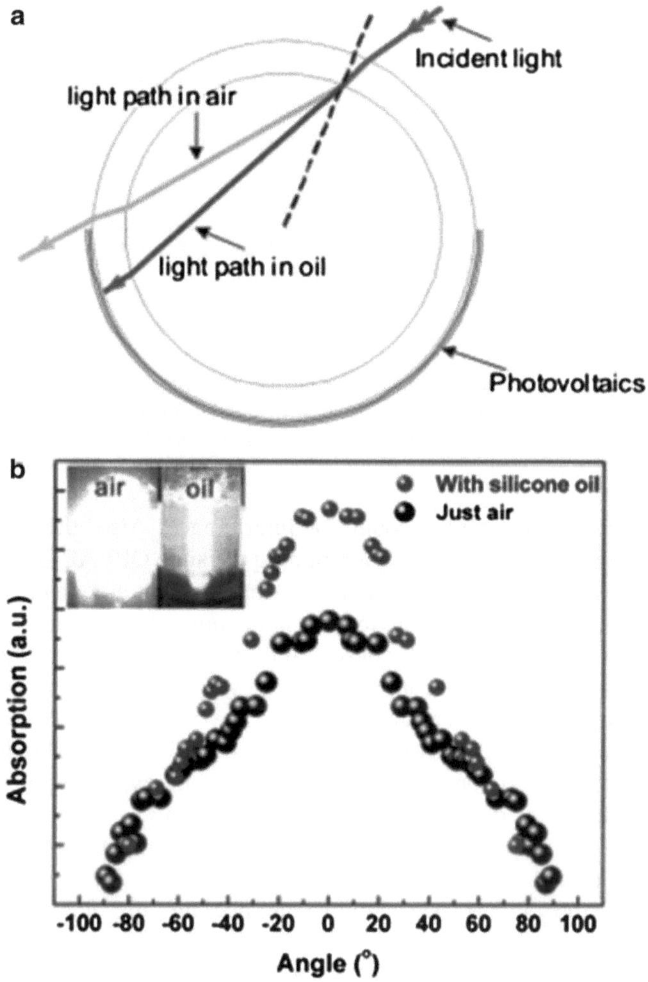

Fig. 7.3 (**a**) Ray diagram in the tube with and without silicone oil. (**b**) Simulation of the absorption of the photovoltaic cell with and without silicone oil. The inset shows the optical images of the tubes with and without silicone oil

the refractive indexes of silicone oil and glass are 1.4 and 1.5, respectively. This simulation is based on two models for two scale calculations: optical path model of the behaviors of reflection and of refraction in tube architecture [18]. Details of the calculation methods used to determine the optical field distribution in thin films are given in the references [19, 20]. Briefly, because the optical paths in tube and silicone oil are of the order of millimeters in our case, we first use ray tracing methods with the Fresnel equations to calculate where the light would occur in reflection and refraction, along with its corresponding angle and intensity. Then, in the thin film (~100 nm), it is necessary to use a Transfer Matrix, to simulate the

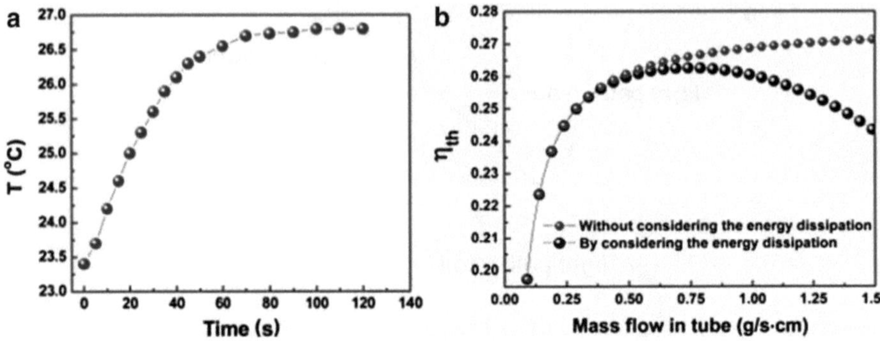

Fig. 7.4 (**a**) Solar thermal characteristics of the device. The temperature rise is measured under AM1.5 g illumination applied normal to the tube surface and is roughly equal for all angles inside of ±25° from 0°. (**b**) The calculated variation in thermal efficiency with silicone oil mass flow rate at the tube

optical field distribution and account for interference. Also, the incident angle dependence was simulated in the programming package OPVAP [21]. The results are shown in Fig. 7.3b, and the simulation is in excellent agreement with the Jsc enhancement observed when the devices are under normal (0°) illumination. Further study is underway to explain the Jsc observed for the oil filled and the air filled tubes when they are under different beam angles; however, we note that the back contacts of the device are added by thermal evaporation and are therefore not continuous at their edges. We therefore expect that little current will be collected there.

Turning to the conversion and collection of heat energy; the temperature of the oil within the tube was measured under static conditions, without the oil flowing. The K-type thermocouple was placed outside of the illuminated area of the device and the temperature was measured at certain time intervals. Again, since the oil was static within the tube, this is an accumulated temperature rise of the silicone oil and is shown in Fig. 7.4.

After about 40s this rate of heat collection "rolls off" as radiative and convective cooling dominates. Under normal operating conditions, this liquid would be shunted into a heat exchanger to maintain a temperature gradient and be replaced by a constant flow of cooled oil. The solar thermal efficiency (η_{th}) could be calculated as Eq. 7.1 [5]:

$$\eta_{th} = \frac{W_u}{G \cdot A_C} = \frac{\Delta Q_u/\Delta t}{G \cdot A_C} = \frac{m \cdot C_p \cdot \Delta T/\Delta t}{G \cdot A_C} = \frac{m \cdot C_p}{G \cdot A_C} \cdot T'(t) \tag{7.1}$$

Where, W_u is the heat collected, G is solar irradiance, C_p is the specific heat of the silicone oil (2.49 KJ/(Kg °C), $A_C$is the collector area (=0.27 cm^2 for one tube cell). When the silicon oil is a circular flow with speed v, its efficiency could be described by following:

$$\eta_{th} = \frac{v \cdot \pi \cdot r^2 \cdot \rho \cdot C_p}{G \cdot A_C} \cdot T'(1/v) \tag{7.2}$$

From this, we could calculate the solar thermal efficiency of the device at different mass flow rates, as Fig. 7.4a shown. Along with the increase of flow speed, its efficiency would be enhanced rapidly, approaching a maximum value. However, for a real industrialized device, the mechanical energy loss, and thermal energy dissipation also needs to be taken into account. If neglecting thermal energy dissipation in transport can be assumed, and only considering the mechanical energy loss of the flowing liquid, that is proportional to the square of flow speed ($\varepsilon \propto v^2$), i.e. this loss is proportional to kinetic energy of flow, we obtain the trend in Fig. 7.4b, which shows an optimum flow velocity to the efficiency peak. It's worth noting that during this operation the photovoltaic part of this device shows good stability since the temperature increase of this tube is relatively very small (3.5°C) [22].

Naturally, we recognized that there are numerous better choices for the thermal fluid (oil); the silicone model system we have chosen provides a nice compromise between heat collection and refractive index for use with the photovoltaic. Alternatively we anticipate that the oil could be loaded with IR collecting dyes to enhance its properties.

To get the total power converted we add the power from photoelectric conversion (η_{el}) to that of photo-thermal conversion (η_{th}):

$$\eta = \eta_{th} + \eta_{el} \tag{7.3}$$

The result is ~28%, though admittedly, the power is in two different forms and thus the heat energy must be extracted using heat exchangers.

In summary, this section demonstrated a PV/T hybrid system based on organic photovoltaics. The tubular geometry, coupled with a high optical index thermal fluid, can be used to concentrate the light on the back photovoltaic, providing enhanced efficiency of photovoltaic conversion and greater daily power generation. The model system we used with P3HT:PCBM, coupled with a silicone oil thermal collector, provides 28% conversion efficiency across a wide spectrum of the sun's radiation. When optimized using the new low bandgap polymers available [23–27] and a better thermal fluid, it is reasonable to expect such devices to exceed 40%, potentially making them a cost effective approach to OPV in general.

7.2 Dye-Sensitized Tube-Based Solar Cell

During the past decade, dye-sensitized solar cells (DSSCs) have attracted much attention because of their short energy pay-back time and facile fabrication process. Its high certified conversion efficiency of 11% reported [28] makes the DSSCs an interesting and promising alternative to conventional silicon solar cells. A typical DSSC is composed of a transparent conductive oxide (TCO) glass such as F-doped SnO_2 glass (FTO) as substrate which, covered by dye sensitized porous nanocrystalline TiO_2 as photoanode, make the inflexible and high weight-volume ratio nature of a FTO glass inconvenient for large area utilization. The planar structure of a DSSC also has lower efficiency of light utilization compare with a 3D structure, due to the dispersion orientations of solar light [29]. A lot of work have been done to enhance the light harvesting by changing the configuration of a DSSC device, such as tandem DSSC [30], bifacial DSSC [31], DNA-like structure [32] or cylinder structure [33] and so on. Among them, the optical fiber based solar cell has been successfully demonstrated in DSSCs [34–36], meaning the optical fiber can been used as a light-trap and optical guider, which can efficiently enhance the solar utilization [37, 38]. Besides that, the fiber cell has four advantages, the first is it has high surface-volume ratio, we can fabricate more power with relative lower sunlight area, which can greatly increase the utilization efficiency of the sunlight; the second is the fiber cell easy realize flexible and knitted; the third is the use of optical fiber instead of metal wire would be especially favorable in harsh environments, where there is a problem of electromagnetic interference, what's more, the corrosiveness of the electrolyte can damage the metal and ruin the device. Finally, it can be applied on the very high light intensity as proved in Sect. 7.3.

Section 5.2 has discussed a glass tube based organic solar cell [39]; the glass tube possess the same properties as an optical fiber. What's more, the tube cell is more conveniently sealed then a planar DSSC structure because it has one end closed structure. Here, we reported a glass tube as light-trap and optical guider based 3-D DSSC (TBDSSC).

7.2.1 Fabrication of Tube-Based DSSC

The tube based DSSC structures were fabricated on glass tubes with one end closed in a hemispherical cap (Chemglass, 0.14 mm inner diameter, 0.25 mm wall, 1.44 mm outer diameter). The design principle is shown in Fig. 7.5a, b. The tubes were cleaned with an ultrasonic bath with ethanol, acetone, and DI water subsequently, and then a 190 nm ITO film was deposited by the magnetron sputtering method on the outer surface of the tube as work electrode. We also deposited ITO film with the same condition on a glass slide to verify the optical and electrical properties of our work electrode. The dense TiO_2 layer (DTL) solution used was the ethanol solution of titanium isopropoxide (0.01 M)/diethanolamine (DEA) (0.02 M). Then, the tube was dipped in the DTL solution for 30s, taken out, and then sintered at 550°C for 15 min. After the sample was cooled to room temperature, a porous TiO_2 layer (PTL) was coated with the dip-coating method in diluted hydrothermal TiO_2 colloid [40] and then was sintered at 500°C for 15 min. The thickness of the PTL was controlled by the TiO_2 colloid's concentration and dip-coating times. Finally, the sintered TiO_2 electrode was rinsed with ethanol and dipped into 0.3mMol N719 cis-di(thiocyanato)-N-N'-bis(2,2'-bipyridyl-4-carboxylic acid-4 -tetrabutylammonium carboxylate) ruthenium (II) and maintained for 12 h at 60°C. The fabricated photoanodes are shown in Fig. 7.5c. For the counter electrode, we rolled up 0.05 mm thick platinum foil like a tube, tightly around the

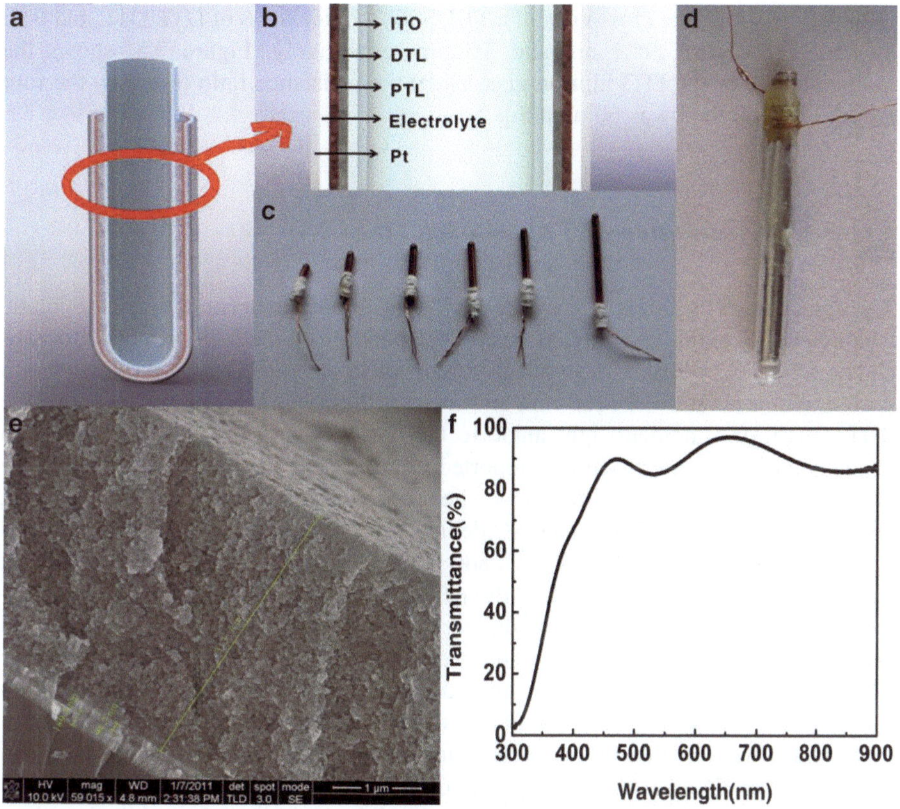

Fig. 7.5 (**a**) Configuration of a tube based dye-sensitized solar cell, the light illuminate at the open end of the tube, (**b**) an enlarged view of the TBDSSC structure, (**c**) as fabricated different length dye sensitized photoanodes, (**d**) optical image of an as fabricated sealed device, (**e**) cross view of scanning electron microscopy (SEM) images of the TBDSSC, (**f**) transmittance of the ITO film deposited on glass slide

PTL working electrode, to form a 3D sandwich structure, then an electrolyte (The electrolyte employed was a solution of 0.6 M BMII, 0.03 M I2, 0.10 M guanidinium thiocyanate and 0.5 M 4-tertbutylpyridine in a mixture of acetonitrile and valeronitrile (volume ratio, 85:15)) was added between Pt foil and the PLT working electrode. Then, the whole assembled cell was sealed into the same kind of tube, with a relevant larger inner diameter. The resulting device photograph image is shown in Fig. 7.5d.

The TBDSSC devices were tested using an AM1.5 g standard Newport No. 96000 Solar Simulator with an illumination intensity of 100 mW/cm^2. The open ends of the tubes were illuminated. Current voltage characteristics were collected using a Keithley 2400 source-measurement unit.

The morphology of the PTL and film thickness of the as-prepared samples was investigated using scanning electron microscopy (Sirion SEM, FEI, Netherlands).

Figure 7.5e shows the cross view of a TBDSSC, the thickness of ITO, DTL and PTL is about 189.4 nm, 94.7 nm and 3.3 μm respectively. Figure 7.5f shows the transmittance of the ITO film on glass slide, the resistance ratio tested by the four point probe method was 60 ohm/Sq.

7.2.2 *Performance of Tube-Based DSSC*

Optimization of the thickness of the PTL is very important because the photovoltaic characteristics exhibit significant variation depending on the thickness of the porous TiO_2. A series of TBDSSCs with different thickness of PTL was fabricated and tested with a length of PTL 6 mm and incident light angle 5°(the influence of the PTL length and incident light angle to the performances of TBDSSC will been discussed later). Same as with the reported stainless steel fiber based DSSC [41], we got the highest conversion efficiency when the thickness of PTL was around 3 μm, as shown in Fig. 7.6a. The TBDSSC device with 3.3 μm PTL was a V_{OC}, J_{SC}, and fill factor of 0.715 V, 4.09 mA/cm^2, and 0.684, respectively, exhibited the highest η, about 2.04%. When the PTL is thinner then 3.3 μm, lower J_{SC} and higher V_{OC} were obtained. That is because the thinner PTL caused the lower dye absorption and lower series resistance, hence caused lower J_{SC} and higher V_{OC}. When the thickness of PTL increased to 5.4 μm, the dye adsorption in the PTL is enhanced and an increased dye-TiO_2 interface would be expected, however, it also caused the larger initial back-transfer rate [41], which induced lower V_{OC} and J_{SC}.

Beside the thickness of PTL, there are also two main factors influencing the performance of a TBDSSC, that is the angle of the incident light and the length of PTL. Because we use open end of tube as light incident path, the incident angle is a definitive factor to the light harvesting of a TBDSSC and hence, influences the efficiency. As we described that, the TBDSSC is a kind of "light trap", once the light incident into tube, it will be totally absorbed by the sensitized PTL after several reflects, so the more light incident into the tube, the more converted photocurrent we can get. As we simulated in Fig. 7.6b, the maximal absorption peak of a device appeared when the incident light angle is around 5° for different length of PTL layer, which is coincident with the conversion efficiency of the devices with different PTL length (Fig. 7.6c). we also can see that, when the PTL length shorter then 5 mm, such as 1.7 mm or 3 mm, the conversion efficiency is a little different with the PTL length longer then 5 mm devices, it shows higher conversion efficiency at smaller incident light angle, that is because at lower incident angle(0° for example), the tube wall will became the optical guider, and the incident light can be absorbed by the dye-sensitized PTL around the tube, according to the Beer–Lambert law [31], a longer PTL length will caused higher series resistance instead of more photocurrent due to the less incident light compare with at 5° incident light angle.

To clearly reveal the variation of the performance of TBDSSCs depend on the length of PTL, we extract device parameters of the TBDSSC with different PTL

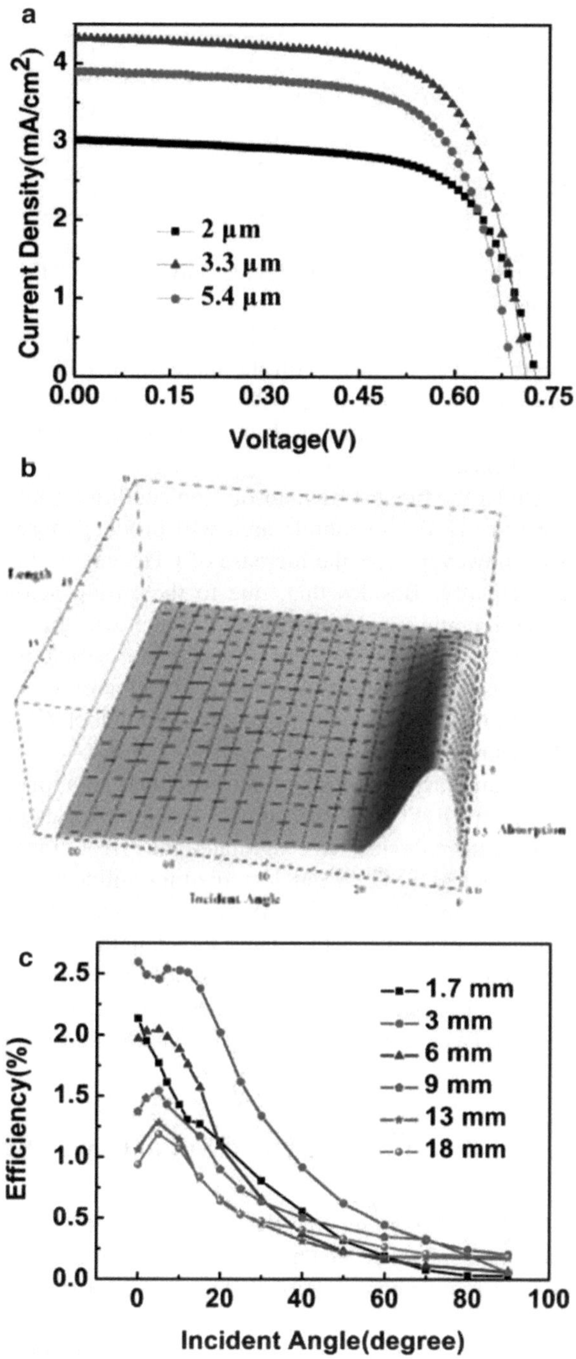

Fig. 7.6 (**a**) J-V curve of TBDSSC with different PTL thickness, (**b**) simulation of the relationship between the PTL length, incident light angle and light absorption. (**c**) The variation of conversion efficiency and length of photoanode based on the angle of incident light

Table 7.1 Performances of Tube-based DSSC

Length	Voc	Jsc	FF	EF
1.5	0.715	3.88	0.625	1.77
3	0.705	5.5	0.624	2.54
6	0.715	4.09	0.684	2.04
9	0.685	3.35	0.647	1.54
13	0.635	3.15	0.626	1.28
17	0.585	3.42	0.564	1.18

length in Table 7.1, all the devices have 3.3 μm PTL thickness and the incident light angle is 5°.

As we can see in Table 7.1, the V_{OC} decrease with the increase of photoanode length, that is because with the increase of PTL length, the photoanode area will increased, hence the longer PTL devices possess larger dye-TiO_2 interface, which caused the larger initial back-transfer rate [41], which is the main cause of the V_{OC} lose, hence induced lower V_{OC}.

Under the biggest light absorption condition (incident angle is 5°), the increased dye sensitized photoanode area will product more photocurrent then the shorter ones, however, with the increase of PTL length, the series resistance will increase dramatically. Besides that, due to the strong acidic nature of the hydrothermal method synthesized TiO_2 colloid, which can lead to corrosion of the Indium tin oxide (ITO) electrode and increase the series resistance, the increased series resistance will evidently counteract the increase of photocurrent. For the 1.7 mm one, too much lower photoanode area cannot convert all the incident light into photocurrent.

In summary, this section demonstrated an approach by using tube-based architecture as light-trap and optical guider to enhance light absorption in DSSC photovoltage devices. We simulated the effect of incident light angle on the absorption of a TBDSSC. We also studied the influence of photoanode length, thickness and incident light angle on the performances of a TBDSSC. We get a maximum efficiency of 2.54% with a 3 mm PTL length, 3.3 μm PTL thickness and 5° incident light angle. The fabricated single device has a high surface-volume ratio, which is promising for the benefits of being flexible and knitted. These results indicate that the tube based DSSC is a promising candidate for high efficiency PV devices.

7.3 Fiber-Based Hybrid Solar Power System

Fiber-based solar cells have proven to have better performance in photonic collection, incident angle dependence, absorption spectrum, and even in thinner active layer to avoid recombination. In this work, it was further proved that fiber-based solar cell may generally work well on very high light intensity (>1,000 Suns), and it can be universally applied on most materials at 10,000 Suns. This could be exceeded by over 50% by further using the Photovoltaic-Thermoelectric Hybrid Concentrating Solar Power System.

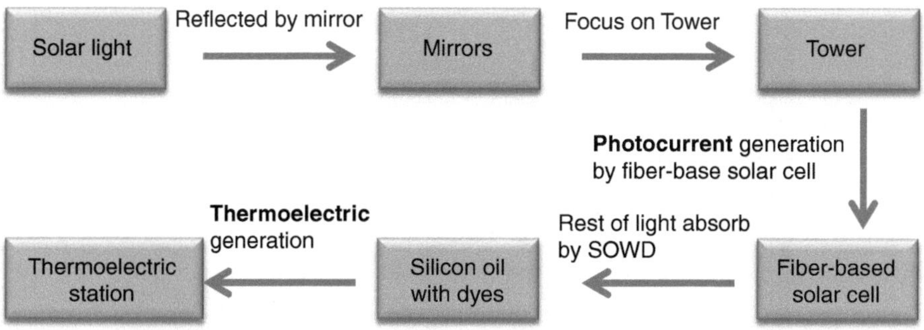

Fig. 7.7 Flow chart of how the FHCSPS works

The recent publication has reported the efficiency would cause decrease in the illumination of high incident light intensity [42]. However, fiber has a large inner area as compared to upper end surface, resulting in the light intensity (incident light/active area) around inner surface being very low and therefore, fiber-based solar cells may be able to perform well on high illumination [43]. This may bring a promising application in the form of concentration on solar system and an outer space solar station near the sun. As we know, the efficiency of solar cell is determined by the product of open circuit voltage (Voc), short current density (Jsc), and fill factor (FF), in which both Voc and Jsc are mainly determined by the band gap. For the solar cells with single active layer, the optimum band gap to efficacy should be around 1.2 eV to transfer 60% photons to current and generate an acceptable voltage, but this still leads to a loss of 40% photons. To make a higher efficiency via balancing Voc and Jsc, tandem architecture was introduced to improve the Voc and keep enough absorption, but this leads to Jsc drop off [44–49] and light loss by the extra layer inserted, such as TiO_2.

For addressing this issue, we developed an optical confinement geometry photovoltaic device, using fiber or tube mostly, to confine light in the cavity and fully absorb light in the range of absorption. By further using the technology of a photovoltaic-thermal hybrid cell [50], we designed a Fiber-based Hybrid Solar Power System (FHSPS), by which the rest of light with lower energy than the absorber's band gap are further absorbed by silicon oil with dye, and this part of thermal energy will be transferred into current in a thermoelectric system. Figure 7.7 shows how the Fiber-based Hybrid Concentrating Solar Power System works.

FHSPS can be applied to some solar system illuminated by high light intensity, such as an Outer Space Solar Power System (OSSPS) near the sun and Concentrating Solar Power Station (CSPS). In Fig. 7.8, solar light focused by mirrors enters into fiber cells from the left side (Red arrows), and then will propagate along the fiber's cavity, coinciding with the current generation on inner surface. After entering into the silicon oil mixed with dyes (Blue arrow), the remaining energy not being absorbed by fiber cells will be fully absorbed as thermal energy. Finally, the solution with thermal power will be used to generate electricity in thermoelectric stations near FHSPS. Figure 7.9 shows some pictures of Concentrating Solar Power Station with FHSPS.

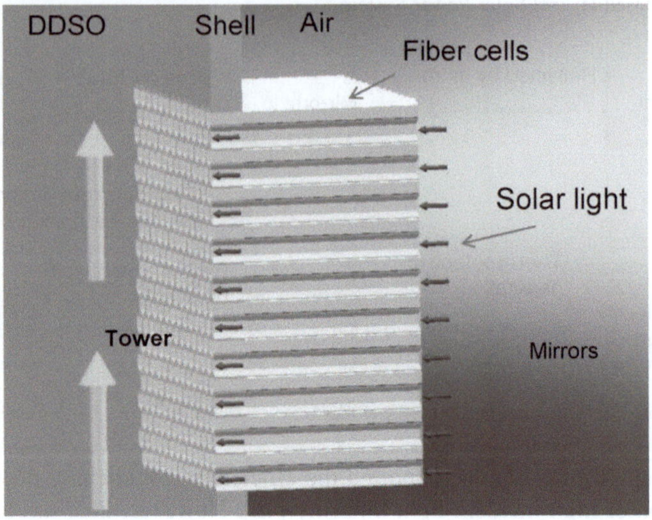

Fig. 7.8 Cross section of fiber-based photovoltaic-thermal hybrid system

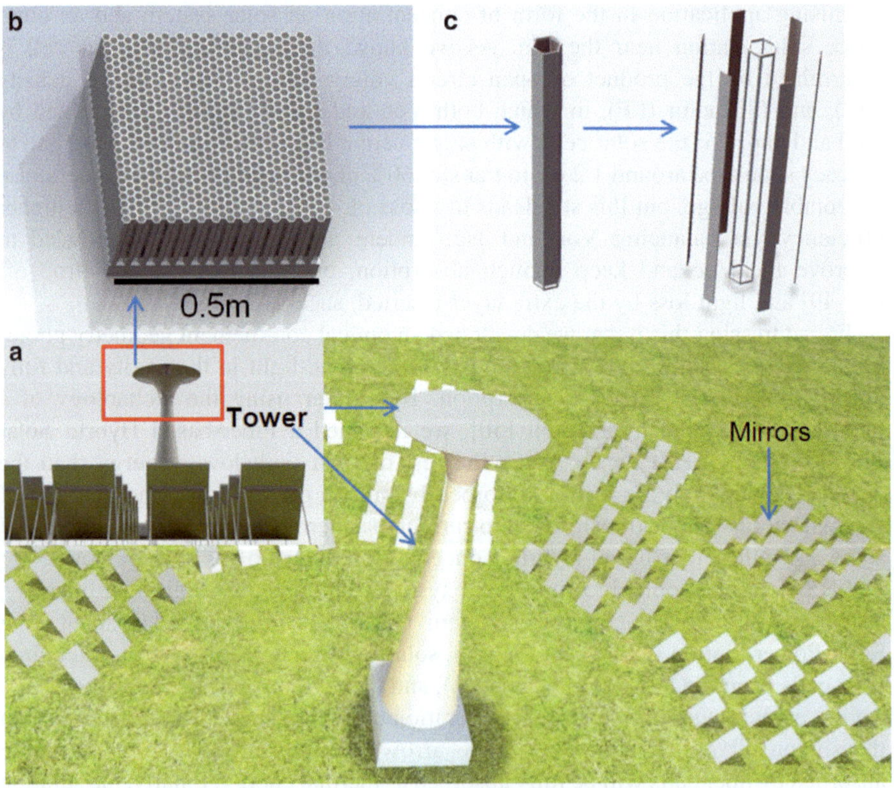

Fig. 7.9 (**a**) The concentrating solar power station with FHSPS at upward view. The inserted figure at *right above corner* is image of up view many fiber-based solar cells are fixed at. (**b**) Is the integrated FBSC matrix. (**c**) Shows how to assemble a FBSC by six slender planar cells

Table 7.2 Efficiencies of concentrating solar power station with FHSPS

	Si	GaInP/GaInAs/Ge	CIGS	CZTS	Polymer	Small molecular
η_{pe} (%)	18.0	43.3	19.7	12.6	9.8	9.5
η_{tte}	16.4	11.34	16.06	17.48	18.04	18.1
η	34.4	54.64	35.76	30.08	27.84	27.6

The photoelectric efficiency η_{pe} predictions based on several materials are listed in Table 7.2. Assuming the thermoelectric efficiency η_{te} could reach 20% in Concentrating Solar Power Station with FHSPS and all thermal could be used out in the process to current, the total efficiency based on Eq. 7.4 for each material could be written as

$$\eta = \eta_{pe} + \eta_{tte} \tag{7.4}$$

$$\eta_{tte} = \left(1 - \eta_{pe}\right)\eta_{te} \tag{7.5}$$

Where, η_{TE} represents the total efficiency (thermoelectric power/total incident power).

Most of the devices have efficiency higher than 30% except two organic cells. Among them, the multifunction GaInP/GaInAs/Ge based on FHSPS performs with the best efficiency over 50%, which is currently the most efficient power station in this world.

References

1. M.A. Green, K. Emery, Y. Hishikawa, W. Warta, Solar cell efficiency tables (version 37). Prog. Photovoltaics **19**, 84–92 (2011)
2. H. Huang, Y. Li, M. Wang, W. Nie, W. Zhou, E.D. Peterson, J. Liu, G. Fang, D.L. Carroll, Photovoltaic-thermal solar energy collectors based on optical tubes. Sol. Energ. **85**, 450–454 (2011)
3. H.P. Garg, R.K. Agarwal, J.C. Joshi, Experimental-study on a hybrid photovoltaic thermal solar water-heater and its performance predictions. Energ, Convers. Manage. **35**, 621–633 (1994)
4. P.A. Davies, A. Luque, Solar thermophotovoltaics: brief review and a new look. Sol. Energ. Mat. Sol. C. **33**, 11–22 (1994)
5. P.G. Charalambous, G.G. Maidment, S.A. Kalogirou, K. Yiakoumetti, Photovoltaic thermal (PV/T) collectors: A review. Appl. Therm. Eng. **27**, 275–286 (2007)
6. T.T. Chow, A review on photovoltaic/thermal hybrid solar technology. Appl. Energ. **87**, 365–379 (2010)
7. S. Kumar, A. Tiwari, An experimental study of hybrid photovoltaic thermal (PV/T)-active solar still. Int. J. Energ. Res. **32**, 847–858 (2008)
8. B. Singh, M.Y. Othman, A review on photovoltaic thermal collectors. J. Renew. Sust. Energ. **1**, 062702 (2009)
9. J.W. Liu, M.A.G. Namboothiry, D.L. Carroll, Optical geometries for fiber-based organic photovoltaics. Appl. Phys. Lett. **90**, 133515 (2007)

10. Y. Li, E.D. Peterson, H.H. Huang, M.J. Wang, D. Xue, W.Y. Nie, W. Zhou, D.L. Carroll, Tube-based geometries for organic photovoltaics. Appl. Phys. Lett. **96**, 243505 (2010)
11. Y. Li, W. Nie, J. Liu, A. Partridge, D.L. Carroll, The optics of organic photovoltaics: Fiber-based devices. IEEE J. Sel. Top. Quant. Electron. **16**(6), 1–11 (2010)
12. N.S. Lewis, Toward cost-effective solar energy use. Science **315**, 798–801 (2007)
13. F.C. Krebs, Fabrication and processing of polymer solar cells: A review of printing and coating techniques. Sol. Energ. Mat. Sol. C. **93**, 394–412 (2009)
14. F.C. Krebs, S.A. Gevorgyan, J. Alstrup, A roll-to-roll process to flexible polymer solar cells: model studies, manufacture and operational stability studies. J. Mater. Chem. **19**, 5442–5451 (2009)
15. F.C. Krebs, T.D. Nielsen, J. Fyenbo, M. Wadstrom, M.S. Pedersen, Manufacture, integration and demonstration of polymer solar cells in a lamp for the "Lighting Africa" initiative. Energ. Environ. Sci. **3**, 512–525 (2010)
16. F.C. Krebs, T. Tromholt, M. Jorgensen, Upscaling of polymer solar cell fabrication using full roll-to-roll processing. Nanoscale **2**, 873–886 (2010)
17. K. Kim, J. Liu, M.A.G. Namboothiry, D.L. Carroll, Roles of donor and acceptor nanodomains in 6% efficient thermally annealed polymer photovoltaics. Appl. Phys. Lett. **90**, 163511 (2007)
18. Y. Li, W. Zhou, D. Xue, J.W. Liu, E.D. Peterson, W.Y. Nie, D.L. Carroll, Origins of performance in fiber-based organic photovoltaics. Appl. Phys. Lett. **95**, 203503 (2009)
19. L.A.A. Pettersson, L.S. Roman, O. Inganas, Modeling photocurrent action spectra of photovoltaic devices based on organic thin films. J. Appl. Phys. **86**, 487–496 (1999)
20. D.W. Sievers, V. Shrotriya, Y. Yang, Modeling optical effects and thickness dependent current in polymer bulk-heterojunction solar cells. J. Appl. Phys. **100**, 114509 (2006)
21. Open Photovoltaics Analysis Platform (OPVAP) by Yuan Li, USA www.OPVAP.com
22. M. Jørgensen, K. Norrman, F.C. Krebs, Stability/degradation of polymer solar cells. Sol. Energ. Mat. Sol. C. **92**, 686–714 (2008)
23. H.Y. Chen, J.H. Hou, S.Q. Zhang, Y.Y. Liang, G.W. Yang, Y. Yang, L.P. Yu, Y. Wu, G. Li, Polymer solar cells with enhanced open-circuit voltage and efficiency. Nat. Photonics **3**, 649–653 (2009)
24. R.C. Coffin, J. Peet, J. Rogers, G.C. Bazan, Streamlined microwave-assisted preparation of narrow-bandgap conjugated polymers for high-performance bulk heterojunction solar cells. Nat. Chem. **1**, 657–661 (2009)
25. X. Gong, M.H. Tong, Y.J. Xia, W.Z. Cai, J.S. Moon, Y. Cao, G. Yu, C.L. Shieh, B. Nilsson, A.J. Heeger, High-detectivity polymer photodetectors with spectral response from 300 nm to 1450 nm. Science **325**, 1665–1667 (2009)
26. L.J. Huo, J.H. Hou, H.Y. Chen, S.Q. Zhang, Y. Jiang, T.L. Chen, Y. Yang, Bandgap and molecular level control of the low-bandgap polymers based on 3,6-dithiophen-2-yl-2,5-dihydropyrrolo[3,4-c]pyrrole-1,4-dione toward highly efficient polymer solar cells. Macromolecules **42**, 6564–6571 (2009)
27. S.H. Park, A. Roy, S. Beaupre, S. Cho, N. Coates, J.S. Moon, D. Moses, M. Leclerc, K. Lee, A.J. Heeger, Bulk heterojunction solar cells with internal quantum efficiency approaching 100%. Nat. Photonics **3**, 297–U295 (2009)
28. Q. Wang, S. Ito, M. Gratzel, F. Fabregat-Santiago, I. Mora-Sero, J. Bisquert, T. Bessho, H. Imai, Characteristics of high efficiency dye-sensitized solar cells. J. Phys. Chem. B **110**, 25210–25221 (2006)
29. K. Uzaki, T. Nishimura, J. Usagawa, S. Hayase, M. Kono, Y. Yamaguchi, Dye-sensitized solar cells consisting of 3D-electrodes—a review: Aiming at high efficiency from the view point of light harvesting and charge collection. J, Sol. Energ. Eng. **132**, 021204 (2010)
30. K. Uzaki, S.S. Pandey, Y. Ogimi, S. Hayase, Tandem dye-sensitized solar cells consisting of nanoporous titania sheet. Jpn. J. Appl. Phys. **49**, 082301 (2010)
31. S. Ito, S.M. Zakeeruddin, P. Comte, P. Liska, D. Kuang, M. Grätzel, Bifacial dye-sensitized solar cells based on an ionic liquid electrolyte. Nat. Photonics **2**, 693–698 (2008)

32. Y. Wang, Y. Liu, H. Yang, H. Wang, H. Shen, M. Li, J. Yan, An investigation of DNA-like structured dye-sensitized solar cells. Curr. Appl. Phys. **10**, 119–123 (2010)
33. Y. Wang, H. Yang, L. Lu, Three-dimensional double deck meshlike dye-sensitized solar cells. J. Appl. Phys. **108**, 064510 (2010)
34. J. Usagawa, S.S. Pandey, S. Hayase, M. Kono, Y. Yamaguchi, Tandem dye-sensitized solar cells fabricated on glass rod without transparent conductive layers. Appl. Phys. Express **2**, 062203 (2009)
35. Z. Tachan, S. Rühle, A. Zaban, Dye-sensitized solar tubes: A new solar cell design for efficient current collection and improved cell sealing. Sol. Energ. Mat. Sol. C. **94**, 317–322 (2010)
36. M. Toivola, M. Ferenets, P. Lund, A. Harlin, Photovoltaic fiber. Thin Solid Films **517**, 2799–2802 (2009)
37. J.W. Liu, M.A.G. Namboothiry, D.L. Carroll, Fiber-based architectures for organic photovoltaics. Appl. Phys. Lett. **90**, 063501 (2007)
38. Y. Li, W. Zhou, D. Xue, J. Liu, E.D. Peterson, W. Nie, D.L. Carroll, Origins of performance in fiber-based organic photovoltaics. Appl. Phys. Lett. **95**, 203503 (2009)
39. Y. Li, E.D. Peterson, H. Huang, M. Wang, D. Xue, W. Nie, W. Zhou, D.L. Carroll, Tube-based geometries for organic photovoltaics. Appl. Phys. Lett. **96**, 243505 (2010)
40. Y. Liu, X. Sun, Q. Tai, H. Hu, B. Chen, N. Huang, B. Sebo, X.Z. Zhao, Efficiency enhancement in dye-sensitized solar cells by interfacial modification of conducting glass/mesoporous TiO_2 using a novel ZnO compact blocking film. J. Power. Sources **196**, 475–481 (2011)
41. X. Fan, Z.Z. Chu, F.Z. Wang, C. Zhang, L. Chen, Y.W. Tang, D.C. Zou, Wire-shaped flexible dye-sensitized solar cells. Adv. Mater. **20**, 592–595 (2008)
42. S.R. Cowan, A. Roy, A.J. Heeger, Recombination in polymer-fullerene bulk heterojunction solar cells. Phys. Rev. B. **82**, 245207 (2010)
43. Y. Li, M. Wang, H. Huang, W. Nie, Q. Li, E.D. Peterson, R. Coffin, G. Fang, D.L. Carroll, Influence on open-circuit voltage by optical heterogeneity in three-dimensional organic photovoltaics. Phys. Rev. B. **84**, 085206 (2011)
44. F. Liu, J.M. Nunzi, Air stable hybrid inverted tandem solar cell design. Appl. Phys. Lett. **99**, 063301 (2011)
45. D.W. Zhao, L. Ke, Y. Li, S.T. Tan, A.K.K. Kyaw, H.V. Demir, X.W. Sun, D.L. Carroll, G.Q. Lo, D.L. Kwong, Optimization of inverted tandem organic solar cells. Sol. Energ. Mat. Sol. C. (2010 in press, Corrected Proof)
46. S. Sista, M.-H. Park, Z. Hong, Y. Wu, J. Hou, W.L. Kwan, G. Li, Y. Yang, Highly efficient tandem polymer photovoltaic cells. Adv. Mater. **22**, 380–383 (2010)
47. X.W. Sun, D.W. Zhao, C.Y. Jiang, A.K.K. Kyaw, G.Q. Lo, D.L. Kwong, Efficient tandem organic solar cells with an Al/MoO(3) intermediate layer. Appl. Phys. Lett. **93**, 083305 (2008)
48. K. Lee, J.Y. Kim, N.E. Coates, D. Moses, T.Q. Nguyen, M. Dante, A.J. Heeger, Efficient tandem polymer solar cells fabricated by all-solution processing. Science **317**, 222–225 (2007)
49. J.Y. Kim, K. Lee, N.E. Coates, D. Moses, T.Q. Nguyen, M. Dante, A.J. Heeger, Efficient tandem polymer solar cells fabricated by all-solution processing. Science **317**, 222–225 (2007)
50. H. Huang, Y. Li, M. Wang, W. Nie, W. Zhou, E.D. Peterson, J. Liu, G. Fang, D.L. Carroll, Photovoltaic-thermal solar energy collectors based on optical tubes. Sol. Energ. **85**, 450–454 (2011)

Chapter 8
Conclusion

8.1 Summary

Optical confinement geometry devices have many distinctive features that can be mainly described by the Table 8.1, and Spectral response, the corresponding simulation methods for which are below.

By means of the theory and prediction introduced above, three series of optical confinement geometry devices are invented and investigated: individual OCGPV, integrated OCGPV, and hybrid OCGPV. Their features are summarized in Table 8.2.

8.2 Suggestions for the Future Investigations

In Sect. 6.4, the ZNR based organic solar cell was studied as an example of near-field approach. In this research, a transfer matrix method was used to calculate the light distribution and corresponding absorption in terms of the complex refractive index of ZNR. This means it is assumed the light distribution in ZNR is one dimensional, i.e. the ZNR was regarded as a successive and even layer instead of discrete rods and gaps. In the next step, a model describing the light behaviors around the ZNR having the dimension closes the light wavelength. In this model, the interference between nano fiber (or tube) should be taken into account. Furthermore, the heterogeneous illumination effect (in Sect. 2.2) should also be brought into near field geometry. Commonly, only one dimensional effect is considered in nanomaterial synthesis or device fabrication. If we could use heterogeneous light to generate another two dimensional force, such as the electrical field between nano fibers, it would be a method to control the behaviors of nanomaterial in micro world.

The second, OCGPV with frequency converter, is another important area for research in the next step. The cross-section for most nonlinear processes is low due to short optical path in planar cells, but OCGPV may perform well since it has a

Y. Li, *Three Dimensional Solar Cells Based on Optical Confinement Geometries*, Springer Theses, DOI 10.1007/978-1-4614-5699-5_8,

Table 8.1 Summary of theories and application to OCGPVs

Theory	Application
Equivalent Circuit	J-V characteristic
	Effect caused by heterogeneous light distribution
	Circuit response at certain illumination intensity
Optical path in cavity	Incident light dependence
	Aspect ratio dependence
	Bottom curvature dependence
Spectral response	Jsc versus thickness of active layer
	Light redistribution in multi-layers
	Absorption and EQE

Table 8.2 Summary of features of OCGPVs

Device	Feature	Deficiency
	Individual OCGPV	
Fiber based solar cell with dip coating ITO	Easy fabrication and low cost, because high vacuum evaporation is not used for preparation	High serial resistance of ITO Rough ITO film
ITO-free Tube based solar cell	No reflection at top end surface of tube	High serial resistance of conductive PEDOT
	Distinctive incident angle dependence	Voc loss
	Cheaper anode as replacement of ITO	
Tandem OCGPV with Serial connection (Parallel connection)	Have higher Voc (Jsc) by tandem subcells Without extra layers between subcells	Low Jsc (Voc) due to limit by subcells
	Integrated OCGPV	
Fiber Bundles	First integrated fibers device	Hard fabrication and film deposition on it
Aligned stamped fiber (dome) devices	Easy fabrication due to no dead corner Avoid Voc loss in fiber cell	Low fill factor
Aligned stamped fiber devices (long)	High absorption Avoid Voc loss in fiber cell	Low fill factor Low voltage Have dead corner
ZnO nano-rod inverted solar cell	Further confine light on the fiber-tip Better incident angle dependence Can combine with nonlinear material Easy fabrication due to no dead corner Avoid Voc loss in fiber cell	
ZnO nano-rod inverted solar cell	High absorption at thicker absorber Micro-fiber cell by near field approach	Low fill factor
	Hybrid OCGPV	
Photovoltaic-thermal solar energy collectors based on optical tubes	Further absorb light to thermal energy	
Dye-sensitized tube-based solar cell	High absorption	Low fill factor
Fiber-based Hybrid Solar Power System	Works well at very high light intensity (>1,000 suns) Broader working range Have universality of most materials	

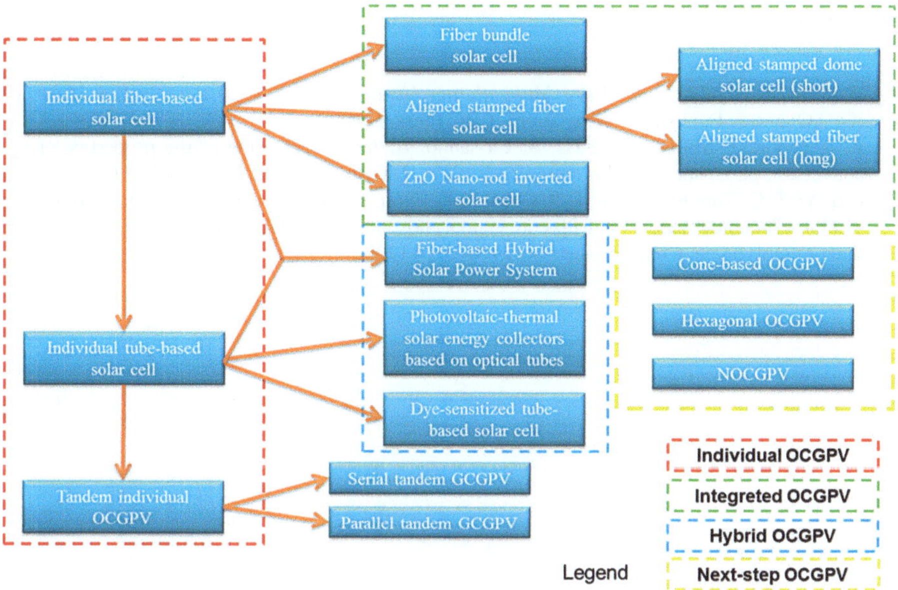

Fig. 8.1 Development history and plan of OCGPV

very long optical path. It is possible to utilize more than one frequency-converting material in these devices. One possibility that will allow the use of the commercial fibers from Kiger is the inclusion of commercial, organically based nonlinear materials such as: H_2C_6N and 4-dialkylamino-1,8-naphthalimides [1]. These materials will be added in the organic layers of the devices, the absorber and the PEDOT buffer, as well will be disposed at interfaces within the device. High-efficiency nanocrystal (~ 10 nm) up-conversion phosphors [2] such as ytterbium and erbium co-doped lanthanum molybdate [3], is another approach to be explored. In addition, compensation for the changing wavelength components along the length can be achieved by inserting frequency shifting of different wavelength bands at different places within the fiber mode structure.

Finally, it has been proven in Sect. 2.3 and Sect. 7.3 that multi-junction OCGPV at very high illumination performs better than other architectures. We have simplified the fabrication process of fiber cells in this concentrating solar system in Fig. 7.9, such that OCGPV could be more easily assembled by several slender planar cells with multijunction architecture. Hopefully, the real device or even this fiber-based hybrid solar power station can be built in the future.

In sum, the three dimensional solar cell by confining light in a geometry is only 4 years old. There are too many issues and problems facing its growth, before maturing into a popular industrial product as the silicon planar solar cell has in the past 50 years. Figure 8.1 shows the development and plan of OCGPV. The red area is individual optical confinement geometry device; Green area is integrated optical confinement geometry device; Blue area is hybrid optical confinement geometry device; and Yellow area the OCGPV device in the next step.

References

1. H.S. Seizo Miyata, *Light Wave Manipulation Using Organic Nonlinear Optical Materials* (OPAOverseas Publishers Association), CRC Press, (2000)
2. A. Rapaport, J. Milliez, M. Bass, A. Cassanho, H. Jenssen, Review of the properties of up-conversion phosphors for new emissive displays. J. Disp. Technol. **2**, 68–78 (2006)
3. G.S. Yi, B.Q. Sun, F.Z. Yang, D.P. Chen, Y.X. Zhou, J. Cheng, Synthesis and characterization of high-efficiency nanocrystal up-conversion phosphors: Ytterbium and erbium codoped lanthanum molybdate. Chem. Mater. **14**, 2910–2914 (2002)

Chapter 9
Appendices

9.1 Instruments and Tools

9.1.1 *Dip Coater*

It is necessary to clean the device (details see Sect. 5.1), and expose the device to ozone for 90 min (rotating the tubes three times every after 30 min) before doing any deposition. As shown in Fig. 9.1, Fiber (As an example of 3D geometry) devices were fixed by Device holder. The step motor at the top of the Dip coater drives the transmission to dip in the solution boat.

9.1.2 *Rotating Deposition*

Fibers are fixed in the holders on the bottom of the auto-rotator. The rotator is driven by AAA batteries sealed in the control box. It can hold 11 fibers or fiber-substrates at the constant speed 360 °/min for more than 72 h without any other power supply. Therefore, the whole deposition includes aluminum, and LiF or other materials can be used without reopening the chamber Fig. 9.2.

9.2 Device Fabrication

9.2.1 *Fiber Preparation*

Mostly multimode solid core fibers were chosen to fabricate the fiber devices. Before device fabrication, the jackets of multimode fibers were stripped off and

Y. Li, *Three Dimensional Solar Cells Based on Optical Confinement Geometries*, Springer Theses, DOI 10.1007/978-1-4614-5699-5_9,

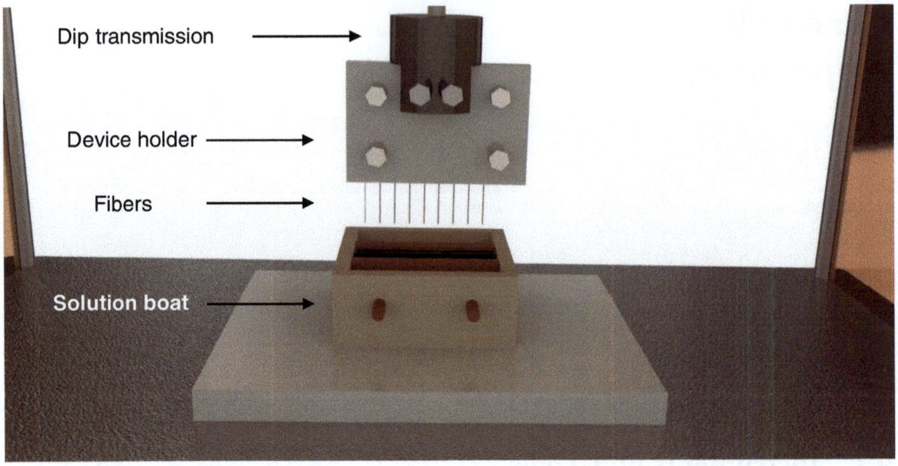

Fig. 9.1 Dip coater

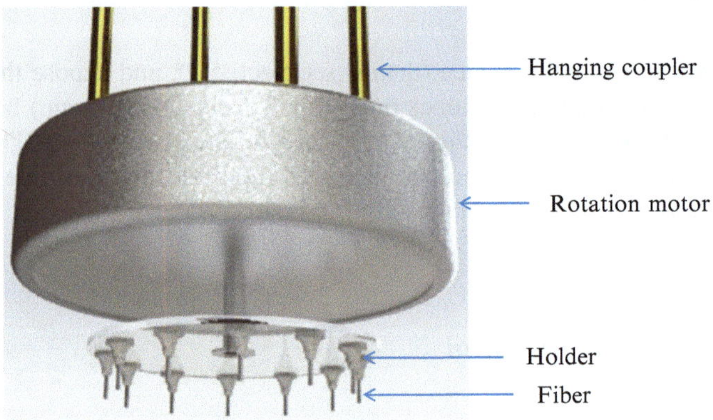

Fig. 9.2 Auto-rotator for high vacuum evaporation

the hard polymer claddings were removed. Then, the core of the fiber was cleaned in an ultrasonic bath with de-ionized water, acetone, and isopropyl alcohol successively for 20 min and dried in an oven for 15 min at 100°C.

9.2.2 ITO Process

9.2.2.1 Sputtering ITO

The ITO films with a thickness of 100 nm were deposited on the samples by a radio frequency (rf) magnetron sputtering (BOSCH) system. During the sputtering, a

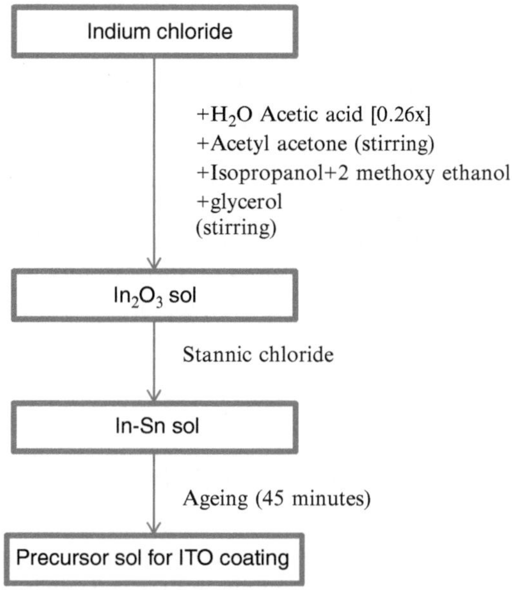

Fig. 9.3 Schematic representation of precursor sol preparation for ITO coatings

99.99% pure ITO ceramic pellet was used as the sputtering target while keeping the flow rate ratio of oxygen and argon, substrate temperature, sputtering power, and chamber pressure at 98:2, 80°C, 90 W, and 1.0*10–5 Pa, respectively. Then the as-prepared samples were annealed in a furnace at 450°C under air atmosphere.

9.2.2.2 Sol-Gel ITO Synthesis and Dip Coating

The sol-gel ITO is synthesized based on the Ref. [1]. The synthetic route is shown in Fig. 9.3

Briefly, ITO sol gel alcoholic solutions have been prepared from indium chloride and stannic chloride with three different ITO concentrations of 6 wt%, 10 wt%, and 20 wt%. The ITO films were prepared on stripped optical fibers by multiple coating. Dip coating was performed at room temperature with fiber withdrawal speed of 8.0 cm/min to form a thin and uniform film [2] using a PCTL0.01 DIP COATING system. After each deposition, the films were heat-treated in air at 500°C for 30 min [3–5]. After multi-layers had been deposited, the films were annealed in N_2/H_2 (1%) atmosphere at 550°C for another hour followed by cooling to room temperature. The linear resistivity was measured by two-point method using a FLUKE 111 multi-meter, which was applicable for the one-dimensional feature of optical fibers. The thickness was determined by SEM image showing the cross-section of each fiber.

9.2.2.3 Remove and Clean ITO

ITO covered glass (purchased from delta R = 8 ~ 10 ohm/sq) was cut into 2.5 cm × 2.5 cm chips. The chips were covered by a strip of acid resistive tape (width ~1.5 cm), then ITO was removed from the exposed area by etching with hydrochloric acid (ACS reagent, 37%, sigma-aldrich). The resulting chip was cleaned in an ultrasonication bath with triton, distilled water, acetone and isopropanol for 15 min each. After drying in a vacuum oven for 1 h, the chips were exposed in ultraviolet ozone plasma for 30 min. A PEDOT:PSS (~40 nm, CLEVIOS PH 500) film was spin-casted at 4,000 rpm on top of the ITO substrates and dried in a vacuum oven for 10 min at 100ºC. These substrates were then transferred into a glovebox for active layer deposition.

9.2.3 Spray Coating Polymer

Spray coating is most commonly applied using commercially available airbrushes wherein a solution coming from a feed cup is transported along a needle to a nozzle by a carrier gas. At the nozzle, droplets form and are ejected out of the airbrush by the carrier gas. The size of the droplets ejected is determined by the nozzle diameter and nozzle-needle distance. Significant solvent is lost in the process of deposition. Ultimately, the deposited film has a morphology that is rough at the micro-scale but is pinhole free and uniform over the whole area. The spray coating process could be considered as a scalable, semi-wet, deposition technique with the possibility of controlling morphology. Such morphological control under droplet drying rates could lead to a balance between re-dissolving and accumulation upon deposition.

The carrier gas was nitrogen (10 psi) and the spray nozzle was held at constant distance of 15 cm from the substrate. The substrate was kept on a hotplate, and its temperature was varied in order to optimize the solar cell performance, as shown in Fig. 9.4.

More details are discussed by Ref. [6].

9.2.4 ZnO Nano-rod Growth

A zinc acetate solution (0.2 mol/L) in methanol was spin coated at 4,000 rpm onto the ITO glass The substrates were subsequently annealed at 400°C for half an hour in air to convert the Zinc acetate to ZnO. From this ZnO-seeding, ZNRs were grown by immersing the substrates in an open bath in 60 ml precursor solution containing 1.2 mM zinc nitrate hydrate, 1.2 mM hexamethylenetetramine at 80°C

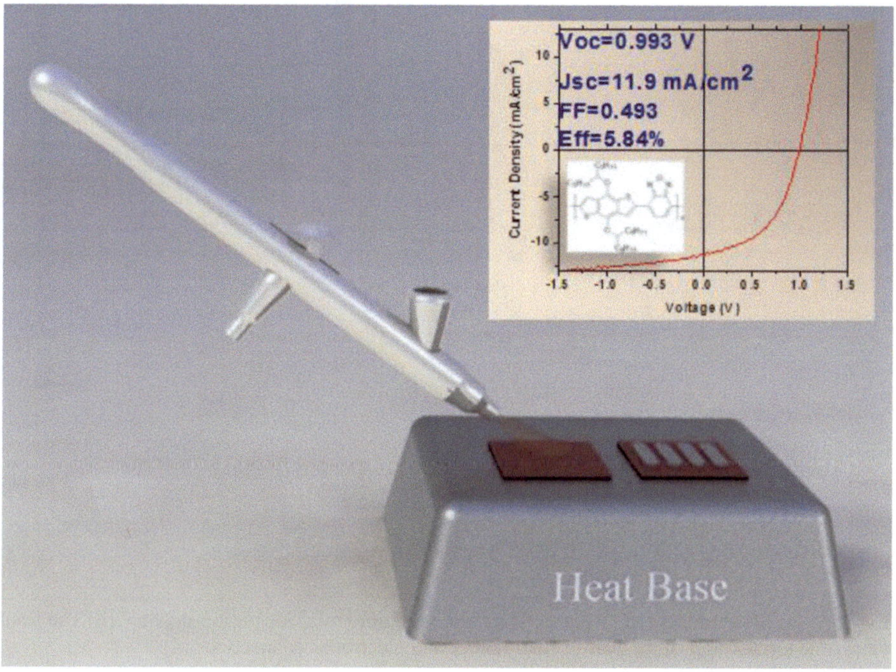

Fig. 9.4 Spray coating procedure. The *right inserted figure* is the IV curve of device fabricated by spray coating using C11-PBDTBO [6]

for 2 h. Afterwards, the samples were taken out and rinsed with de-ionized water (ultrasonic bath for 3 min) and baked in air at 100°C for 30 min to remove any residual water.

9.3 Testing and Measurement

- **Current-voltage measurement**

Current voltage characteristics were collected using Keithley 236 source-measurement unit using an AM1.5 g standard Newport # 96000 Solar Simulator, with a 150 W 6255 Ozone Free Lamp and an AM1.5 global filter Newport # 81094. The solar mismatch is 0.18 when operating with an illumination intensity of 100 mW/cm2.

As shown in Fig. 9.5, the front of the fiber is apertured by fitting the fiber end through a hole in an optical baffle with a diameter equal to the fiber. Thus, no light was incident on the fiber's side and the overall angle of incidence was 90° to the face of the fiber.

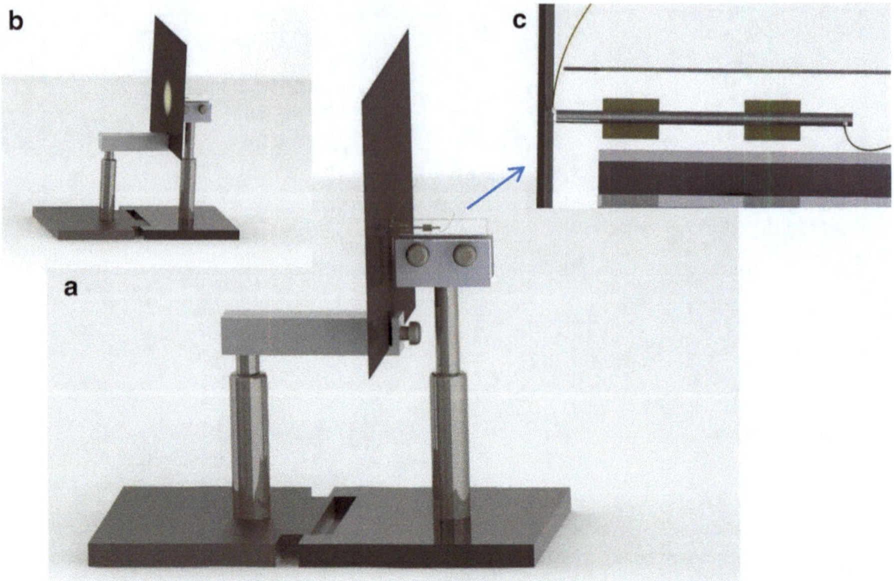

Fig. 9.5 (**a**) Measurement of IV curve. The fiber is aperture by black plastic shading. (**b**) The front view of (**a**). (**c**) Zoom in the part of fiber holder and electrodes connection

- **Absorption and Transmission**

The absorption and transmittance spectra of all films were characterized by an UV–Vis Spectrophotometer Varian, Cary 50.

- **External Quantum Efficiency (EQE) measurement**

The external quantum efficiency (EQE) was measured using a Newport Cornerstone 260 Monochromator in conjunction with a Newport 300 W Xenon light source. The incoming light was chopped at 30 Hz and measurements were taken by a Newport Merlin # Lock-In amplifier. These measurements were compared to a Newport # calibrated photodiode array to determine EQE.

- **Scanning Electron Microscopy**

The thickness and grain structure was determined by FE-SEM imaging of fractured fiber cross-section.

- **Atomic Force Microscopy**

Tapping mode AFM images were taken by JEOL JSPM-5200 atomic force microscope Fig. 9.6.

- **Heterogeneous light illumination generator**

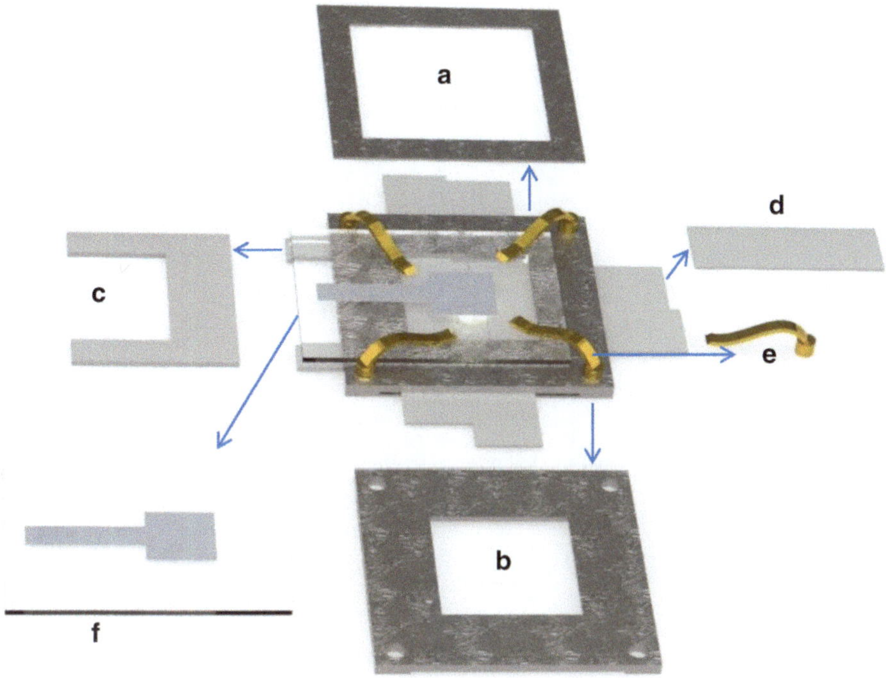

Fig. 9.6 Heterogeneous light illumination generator. (Refer to Fig. 2.6) Parts and its functions are marked as: (**a**) base: supporting whole tool; (**b**) cover: fixing light barrier; (**c**) and (**d**) light barrier: building an aperture with changeable shape; (**e**) clip: fixing the sample; (**f**) sample of solar cell

References

1. S.R. Ramanan, Dip coated ITO thin-films through sol-gel process using metal salts. Thin Solid Films **389**, 207–212 (2001)
2. C.J. Brinker, G.C. Frye, A.J. Hurd, C.S. Ashley, Fundamentals of sol-gel dip coating. Thin Solid Films **201**, 97–108 (1991)
3. K. Daoudi, B. Canut, M.G. Blanchin, C.S. Sandu, V.S. Teodorescu, J.A. Roger, Densification of In2O3: Sn multilayered films elaborated by the dip-coating sol-gel route. Thin Solid Films **445**, 20–25 (2003)
4. M.J. Alam, D.C. Cameron, Investigation of annealing effects on sol–gel deposited indium tin oxide thin films in different atmospheres. Thin Solid Films **420**, 76–82 (2002)
5. A. Beaurain, D. Luxembourg, C. Dufour, V. Koncar, B. Capoen, M. Bouazaoui, Effects of annealing temperature and heat-treatment duration on electrical properties of sol–gel derived indium-tin-oxide thin films. Thin Solid Films **516**, 4102–4106 (2008)
6. W. Nie, C.M. MacNeill, Y. Li, R.E. Noftle, D.L. Carroll, R.C. Coffin, A soluble high molecular weight copolymer of benzo[1,2-b:4,5-b′]dithiophene and benzoxadiazole for efficient organic photovoltaics. Macromol. Rapid. Comm. **32**, 1163–1168 (2011)

About Author

Yuan Li (李远) was born in Beijing, China, on February 25, 1983. He received his B.S. from Beijing Jiaotong University in 2005. He was then recommended for admission to the Institute of Optoelectronics Technology, Beijing Jiaotong University, to pursue his M.S. During his research, he focused on the project "Behaviors of electrons in organic-inorganic complex electroluminescence device," which was supervised by Prof. Xurong Xu. In August 2008, he received an offer from the Department of Physics, Wake Forest University, USA, to work as a doctoral student. His Ph.D. thesis focused mainly on the device physics of three-dimensional solar cells, especially for fiber-based geometries (more details are available at www.ly.inwake.com).

Y. Li, *Three Dimensional Solar Cells Based on Optical Confinement Geometries*, Springer Theses, DOI 10.1007/978-1-4614-5699-5,

MIX
Papier aus verantwortungsvollen Quellen
Paper from responsible sources
FSC® C105338

If you have any concerns about our products,
you can contact us on
ProductSafety@springernature.com

In case Publisher is established outside the EU,
the EU authorized representative is:
Springer Nature Customer Service Center GmbH
Europaplatz 3, 69115 Heidelberg, Germany

Printed by Libri Plureos GmbH
in Hamburg, Germany